A Biologist's Guide to a Restorative Home

Katherine T. Fuller
B.S. (Biology) M.S. (Ecology)

Table of Contents

Purpose

A key tenet of the science of biology is maintaining homeostasis or balance in one's body. A synonymous concept, sustainably, infers living in balance with nature and one's environment. A sustainable lifestyle promotes each of us to make conscious efforts to minimize disruption to homeostatic balance while simultaneously minimizing our impact on the environment. Our challenge, once presented with all the conveniences of modern living, how do we discern, choose, and then internalize habits to better achieve a more sustainable lifestyle?

The lifelong benefits of a sustainable lifestyle are tangible. Benefits, such as: improved physical and mental health, having greater vitality, retaining the ability to make decisions independently, increased self-sufficiency; are all achievable while fostering within us a sense of personal commitment and contentment. Having degrees in biology and ecology have taught me we have a wondrous capacity to maintain homeostasis within ourselves and with our environment. However, this capacity is diminished by making non-supportive choices. The good news, we can replenish this wondrous capacity by taking a different path, one of choosing a more supportive lifestyle and by doing so, balance is restored.

Prologue

My youth was one of the 60's and what a decade it was! It was quite frankly, a decade of cultural and political upheaval. My impressionable nature was impacted and molded by experiences, such as the collective assassinations of President Kennedy, Bobby Kennedy, and Dr. Martin Luther King. Dr. King was assassinated while I sat at my kitchen table doing homework. I lived a mere 80 miles away. As violence and race riots ensued around the country, I thought this could not be very good. The Viet Nam war and the draft, George Wallace and the KKK, the gruesome Manson murders in LA, and most personally, busing and desegregation of schools presented a tumultuous list of impressionable events. 'This could not be very good' remained a mental constant.

The 1960's also was a chaotic time for the environment. From television news casts, one could see smog enveloping and hanging over western cities like a yellow dome and I thought this could not be very good. Having grown up in the southern mountains, I had seen what remains of the earth, after coal has been stripped from the land. I had seen scars on the mountain sides left after a forest clear cut and I thought this could not be very good. Then, both news and print sources began to inform us of a community in Buffalo, NY. There were reports of women having too many miscarriages, children being born with birth defects, and cancer cases, so numerous, they were termed cancer clusters. Once again, I thought this could not be very good.

Turning the page to the 1970's, the community of Love Canal stayed in the news. Love Canal, once envisioned as the new wave for planned communities, had become an unmitigated environmental tragedy. Before the community was developed, Hooker Chemicals had dumped metal drums of chemical byproducts into the canal's earth. These were byproducts of the manufacture of dyes, perfumes, solvents, and synthetic resins. The drums were buried 25 feet below the ground's surface and then capped with clay to prevent leakage of the toxic contents. Hooker sold the canal dump site to the Niagara Falls School Board in the 1950's for $1.00. Subsequently, two schools with their playgrounds were built over and above barrels of what were now seeping, oozing toxic wastes. Construction of streets, homes, sewer lines, and water lines further contributed to the puncturing of a once protective clay cap. In total, 800 private homes and 240 low-income apartments with their residents, unwittingly became victims of putrid chemicals bubbling up from the ground below their feet (Beck, 1979).

Forty years later, my son and I decided to visit what remained of this once thriving community. Designated as the first EPA Superfund site, its remediation spanned 21 years of time. Our entry was not possible, as the acreage, abandoned streets, demolis

This Photo by Unknown Author is licensed under CC BY-SA

hed homes, businesses, schools, parks, and playgrounds were cordoned off by an 8-foot-high fence and concrete street blockades. No public entry allowed.

Introduction

Native Chemistry versus Synthetic Chemistry

Prior to the 20th century, words like polyethylene, polystyrene, and polytetrafluoroethylene and chemical categories such as PFOAs and PBDEs did not exist. In contrast, these new synthetic manmade chemical creations presented with complex degradation, if at all. **Persistent, forever chemicals**, and **pervasive, everywhere chemicals**, are the present-day challenge of our own making. Indeed, with the evolution of synthetic chemistry, we have designed, created, and manufactured a more toxic, less safe environment.

It is appropriate to differentiate between naturally occurring chemicals and synthetic/manmade chemicals. Organisms exist as organic (carbon-based) biochemical repositories. Our survival depends upon environmental inorganic and organic chemicals continually cycling through organisms utilizing processes such as ingestion, respiration, excretion, and decomposition. Organisms do not survive without the air's oxygen that we breathe or the water (H_2O) that we drink or the food that we eat; life is chemical-based. While it is misleading and inappropriate to promote chemical-free environments; it is appropriate to promote home environments to be toxin-free. The Environmental Protection Agency (EPA) defines a toxic chemical or toxin as any substance which may be harmful to the environment or to one's health if inhaled, ingested, or absorbed across the skin.

Today, nearly everything in our home is made of synthetic materials from walls, to floors, to ceilings, to furniture, and then to what **we as consumers bring into the home**. It is estimated by the American Chemical Council (ACC) that synthetic chemicals are used in 96 percent of manufactured materials and products. Consequently, we have not only created a problem in sheer volume of chemicals produced, but also in their diversity and toxicity.

The Toxic Substances Control Act (TSCA) was passed in 1976 to evaluate the volume and diversity of synthetic chemicals. The EPA was tasked to retrieve, maintain, and continually revise the chemical inventory which roughly stands at 80,000-plus commercial synthetic chemicals just produced in the U.S.A. Safety testing has occurred only with approximately one percent of these synthetic chemicals. Additionally, the TSCA did not provide the EPA with the teeth to fully achieve the bill's goal, as the chemical industry and manufacturers were provided "an out" with a trade secrets loophole. A trade secret includes the formulation, methodology, technique, and process providing independent economic value, actual or potential, from not being known. By virtue of the trade secrets loophole, 17,000-plus secret compounds exist in commercial products (Wambolt, 2021). Its presence has promoted our unintentional exposure to approximately 3,000 synthetic chemicals by absorption across the skin, inhalation, and ingestion over the course of one's lifetime.

Besides the complexity of their chemistry, synthetics present another more vexing and hazardous problem. Native organic and inorganic chemicals break down naturally utilizing enzymes in organisms and cycling processes in the environment. In contrast, synthetic chemicals present a riddance problem. Some synthetic chemicals, such as phthalates, ubiquitous because of their abundant presence and use, are metabolized well by the body. Biologically, phthalates exhibit short half-lives as they decompose through the body's liver detoxification pathways and we excrete them from the body in urine and in the feces (Wang, 2021). However, phthalates are so abundant in consumer and household products; they are termed **pervasive or everywhere chemicals**. There is 'no away' for us from their toxic presence and exposure.

Alternatively, other synthetics are not well metabolized by the body. **Persistent synthetic chemicals or forever chemicals**, are chemically quite stable, resisting biological breakdown; thus, they stay in place in organisms and the environment. Because we do not possess the excretory means to address their breakdown, we retain them as chemical residues in our bodies. For example, one class of everywhere synthetic chemicals, PBDEs or polybrominated diphenyl ethers were formulated as flame/fire retardants. These compounds bind with the foam materials making up upholstered furniture, bedding, household carpeting, and drapery fabrics. However, with every sitting down or standing up, some of the retardant chemicals puff out, residing as unwanted toxic components of household dust. Living in our chemical-based, modern, industrialized society and home has placed each of us at risk of toxic exposure. According to the Centers for Disease Control (CDC) each of us carries within our bodies approximately 200 chemical residues, termed the body burden.

Persistent, non-biodegradable synthetic chemicals, such as PBDEs, hitchhike in household dust where they are available to be inadvertently ingested, inhaled, or absorbed across the skin. Inside the body, non-metabolized, fat-soluble toxic residues can bind with fatty tissues in a process termed **bioaccumulation**. Since humans exist at the top of the food chain, upon death and decomposition the toxin will be released back into the environment. Alternatively, toxically tainted organisms lower on the food chain are consumed by other organisms precipitating the process of **biomagnification,** allowing toxin levels to expand within the greater environment.

The presence of synthetic chemicals within our bodies is a given. Their burden is indeterminate, because factors such as genetics, metabolism and weight, nutrition, and other environmental exposures, play a role in their manifestation as toxins within the body. Being aware of their presence warrants how best to address them.

Present day health challenges see us facing increased incidence of chronic disease because of toxic exposures. Researchers at the University of California (Davis) tested 200 toddlers and found 96 chemicals residing in their bodies. Many of the chemicals were linked to hormone and brain disruption (Oh, 2025). Toxic exposures are suspected with increased incidences of childhood cancers (Ward, 2014). A joint study between the University of Vermont and Denmark's Aarhus University, found childhood phthalate exposure was strongly associated with the incidence of osteosarcoma (bone cancer). In fact, exposure to endocrine-disrupting (EDC) phthalates was associated with a 20 percent higher incidence hazard for any childhood cancer (Ahern, 2022).

Harvard's T.H. Chan's School of Public Health reports that the incidence of miscarriage is increasing by roughly one percent each year. It is hypothesized that female exposure to phthalates increases the risk for gestational diabetes which correlates with an increased risk of miscarriage (Feldscher, 2016). Fetal development is affected by phthalates, especially language and motor skill development, which is not expressed until after the baby is born. Phthalate exposure is also suspected to cause reproductive issues in males and females resulting in reduced fecundity or potential to produce offspring (Green, 2021). It is most unfortunate that phthalates as plasticizers exist in so many different household products such as, vinyl flooring, cosmetics, hair and body care products, household cleaners, even the food we eat (Westervelt, 2015).

Many compounds act as hormone mimics with endocrine disruption the result. Identified as **EDCs**, these **endocrine disrupting compounds** or **endocrine disruptors** are chemicals which commandeer the production, release, transport, metabolism, or binding of hormones at their target sites in the body. Cell receptors designed to receive hormonal signals to initiate changes in the body may be promiscuous, allowing an EDC to dock instead of the native hormone; thus, cell communication is disrupted.

Researchers measuring indoor air quality of homes in Massachusetts found, on average, 19 hormone disruptors present in retrieved air samples and 24 hormone disruptors bound in household dust samples (Rudell, 2010). It bears repeating, living in our chemical-based, modern, industrialized society and home has placed each of us at greater risk of toxic exposure. A study performed in 2005 found the presence of 200-plus synthetic chemicals in the umbilical cord blood of the 10 babies tested. Thus, the babies were born pre-polluted. Studies indicate the air inside our homes is 2 to 5 times more polluted than outdoor air because of the presence of synthetic chemicals and most of us spend 90 percent of our time indoors (Wambolt, 2021).

It is not as if the public has been kept in the dark about the problem of chemical synthetics and their toxic nature. Concepts, such as greening one's home and greenwashing, were introduced into the vernacular long ago. In fact, companies still employ greenwashing tactics utilizing subliminal messaging by selling products in green bottles or utilizing green labels on packaging. Greenwashing causes consumers to assume that what the company produces and sells is made with the consumer's environmental conscience, and personal health in mind. This has now stimulated the growth of novel eco-friendly or nature-based companies which have moved beyond household cleaning products, into the beauty arena, and products specifically earmarked for babies and infants. A similar marketing strategy arose with fabrication of unscented products, to fool consumers into believing if the product does not have a scent, then it must be harm-free. Companies are in the market to sell their products. Consequently, they employ marketing tactics to embellish their care for you, the buyer.

The Environmental Working Group (EWG) performed a 14-month investigation on chemical toxins present in popular cleaning products, in which some of them were advertised as "green." Simple Green All-Purpose Cleaner is labeled non-toxic and biodegradable; yet, this "green" cleaner contains 2-butoxyethanol, an eye irritant solvent absorbable through the skin that damages red blood cells. Spic and Span Multi-Surface Cleaner containing nonylphenol ethoxylate, decomposes to form nonylphenol, a recognized hormone disruptor. Scrubbing Bubbles-Antibacterial Bathroom Cleaner contains up to 10 percent butoxydiglycol, a solvent that irritates and causes inflammation of air passageways. Mop and Glo Multi-Surface Floor Cleaner contains methoxydiglycol, another suspected endocrine disruptor. Chlorox, Fantastik, Febreeze, Formula 409, Easy-Off, Lysol, Mr. Clean, and Spic and Span are laced with ethanolamine, classified as an asthmagen or asthma trigger, by the Association of Occupational and Environmental Clinics. Febreeze Air Effects, an air freshener, emits 89 different air contaminants. Synthetic air fresheners do nothing to freshen or clean the air; instead, they act to pollute it. The most damning aspect of the report, just 7 percent of the cleaning products evaluated adequately disclosed their contents; while 53 percent contained respiratory irritants and carcinogens, such as formaldehyde and chloroform ("EWG Cleaners Hall of Shame Reveals Hidden Hazards," 2012).

In contrast, sustainability or sustainable practices should be the mantra instead of the greening concept, now passé. Ecological sustainability predicates not doing harm to the environment and utilizes practices and processes which conserve, not deplete required resources. Living sustainably infers living in balance with nature and one's environment (homeostasis), a 20th century model from long ago, whose time must come again. We should embrace analogous measures to maintain homeostasis within our bodies and this begins with the choices and actions we take within our homes. Our very survival, as living organisms, depends upon adoption of a more sustainable and toxin-free lifestyle to the best of our ability. We opened Pandora's Box with the introduction of thousands of synthetic chemicals; now our challenge becomes, **how do we reduce the risks?**

Our Toxic Household (Phthalates)

What comprises our list of household toxic exposures? As consumers, we introduce many toxic exposures via commonly purchased household products. Household cleaning products, personal care products, various detergents, household solvents, air fresheners, and cosmetics all expose us to a daily toxic regimen. Therefore, we accrue a lifetime of toxic exposures which depending upon our body's ability to degrade or not degrade them, leads to bioaccumulation of chemical toxic residues residing within body tissues.

Fragrance often denotes a connotation of sweetness or delicacy. However, beware of fragrance's false witness, since it indicates the presence of hidden toxins. Fragrance is listed as ingredient on labels of air fresheners, scented candles, furniture polish, antibacterial soaps, hand sanitizers, shampoos, body gels, lotions, deodorants, cleaning products, laundry and dishwashing detergents, dryer sheets, carpet cleaners, personal care and beauty products, perfumes, and colognes, etc. In fact, fragrance is the fastest-growing entity within the booming beauty industry, out pacing the creation of new makeups, skin, and hair care products. Even products advertised as unscented still may list fragrance as an ingredient with hundreds of substances hidden and protected.

Fragrance as a chemical ingredient in so many consumer products is sometimes listed and sometimes not. Whether listed or not, fragrance additions hide the presence of **phthalate** compounds. The presence of phthalates in our consumer products is so prevalent, there is simply no way for the consumer to avoid their presence. As a chemical group, phthalates stand as the poster child illustrating the concept of pervasive or everywhere chemicals. To illustrate, the Environmental Working Group (EWG) found more than 75 percent of products tested listing fragrance as an ingredient, contained phthalates. Phthalates act as chemical fixatives or as chemical plasticizers. Fragrance demands phthalate fixatives, so that the odor is better retained by the product and once the odor is emitted, the odor chemically lingers in the air (Houlihan, 2002). One becomes a wiser, more health-conscious consumer with this awareness. Avoid synthetically fragranced consumer products.

As plasticizers, phthalates are added to plastics to make them more malleable or flexible. Inventory the amount of plastic in your home and then add to that inventory, the number of scented or fragranced products and you quickly understand pervasive/everywhere. Besides fragranced products, we are exposed to phthalates from anything that is vinyl including: flooring, adhesives, lubricants, tubing, plastic clothes, shower curtains, inflatable toys, garden hoses, food containers, plastic food packaging, even medicinal time-released capsules. For example, some anti-inflammatory drugs and antibiotics which require extended or delayed release in the body do so with the presence of phthalates. Phthalates leach into foods from processed food packaging and from plastic microwavable containers. Phthalates are problematic, not because we store and accumulate them within our tissues, but because our daily exposure is so collectively overwhelming. Even though we readily metabolize them via excretory pathways associated with the liver, the overabundance of phthalate exposures overwhelms degradation pathways. Phthalates enter our bodies through a plethora of consumer products including household cleaners and personal care products, and with any spritz of a fragranced product. Therefore, we must alter and reduce daily toxic phthalate overload with a change in our household consumer product choices to better support optimal function of degradation pathways.

Phthalates, as EDCs, wreak havoc on the endocrine messaging system. The action of endocrine disruptors in the body threatens human health and wellness. Consequences include: alterations to sperm quality and fertility in males, abnormalities of the sex organs, endometriosis, precocious puberty, altered nervous, immune, and respiratory function, metabolic issues, type 2 diabetes, obesity, cardiovascular issues, certain cancers, and neurological and learning disabilities (Arrigo, 2023). We are exposed to phthalates in a wide-range of fragranced and unscented consumer products which we use throughout the day. Just reflect on your daily use of shampoo and body wash, lotions, facial and hair care products, air fresheners, drier sheets, laundry and dishwashing detergents, other cleaning products, cosmetics, and perfumes. Illustrating one exposure, a survey of 406 men found those who topically applied cologne or aftershave products presented much higher urinary levels of phthalate derivatives compared to men who did not use such products (Duty, 2005).

It is far more concerning to evaluate the effects of phthalate exposure on a developing fetus, a newborn, and an infant? *In utero* exposure occurs during critical periods of fetal neurodevelopment which leads to harm that can never be undone. Dr. Shanna Swan, a reproductive epidemiologist writes, *"Hormone interfering substances, such as phthalates, interfere with normal function, including testosterone and estrogen. Even in small doses, they pose harm to unborn babies and young children whose bodies are growing rapidly. These are hormone-warping chemicals, which can enter the placenta, having the ability to alter the anatomical development of girls and boys, change brain function, and impair the immune system."* (Swan, 2021).

Each of us is endowed with our own unique set of DNA/genes/chromosomes which we received from our two parents. Our genes are not expressed, turned on or off, all at the same time; synonymous with turning the various light switches on or off in one's home. Epigenetics is all about gene control or gene regulation and we want our DNA on/off signals to function correctly and at the proper time. Can environmental exposures from our food supply, alcohol and drugs, or environmental toxins (phthalates and other EDCs) alter these signals? Yes, scientists recognize that exposure to EDCs during fetal development can alter epigenetic signaling processes. Epigenetic modifications during embryonic development, *in utero*, may present long term health effects, such as impaired development, reproductive complications, heart issues and an increased risk of neurological disorders (Akanbi, 2025).

Research readily supports phthalates move across the placental barrier in expectant mothers. More than 90 percent of babies in one study showed the presence of the breakdown of diethyl phthalates (DEP) in urine samples. Some human studies suggest *in utero* phthalate exposure may lead to abnormal genital development in males with an increased risk of hypospadias, undescended testes, and increased incidence of male infertility due to sperm damage (Hu, 2009). Dr. Philip Landrigan, an expert on EDCs, associates fetal exposure to phthalates as a contributor to the increased incidence of childhood neurological disorders, such as autism (Reynolds, 2025). Studies conducted during the last decade support associations exist between prenatal phthalate exposure and the incidence of attention-deficit hyperactivity disorder (ADHD), aggression, defiance, emotional reactivity, delinquent behaviors, adverse cognitive development including lower IQ, reduced psychomotor development, and impaired social communication (Engel, 2021).

In infants, toddlers, children, and adolescents, phthalate exposure aligns with developmental milestones. Crawling babies are in greater contact with vinyl floor products, hand to mouth contact made with phthalate-laced plastics, and contact via phthalate-laced household dust. Infants and toddlers are at particular risk because of their higher intake of toxins per unit of body mass and their elevated metabolic and respiratory rates (ATSDR, 2023). In other words, little people are more intimate with their environment, physically and metabolically. Their intimacy presents an immediate risk for skin and respiratory allergies from toxic-laden dust. However, it is the long term, collective and accumulated exposures which present a more dramatic and harmful outcome.

Indeed, a lifetime of exposure to phthalates predisposes us to premature mortality with increased risk of developing obesity, type 2 diabetes, and cardiovascular disease. A study performed by Dr. Leonardo Trasande and his researchers at Grossman School of Medicine (NYU), studied 5,000 American adults between the ages of 55 to 64. The study found that those adults showing the highest concentrations of urinary phthalate derivatives were more likely to die of heart disease than those having reduced levels of phthalate derivatives. Phthalates are recognized to disrupt testosterone and reduced levels of testosterone is a predictor for the onset of adult cardiovascular disease. Furthermore, the study estimated 91,000-plus premature deaths may occur each year for Americans in this age group. Dr. Trasande writes, *"It's clear that limiting exposure to toxic phthalates can help safeguard Americans' physical well-being."* ("Deaths linked to 'hormone disruptor' chemical," 2021).

A study, recently published in *eBiomedicine*, examined the impact of Di(2-ethylhexyl) phthalate (DEHP) on global populations. The team at Grossman School of Medicine (NYU) analyzed health and environmental data from population surveys which included urine samples containing the chemical derivatives left from the breakdown of DEHP. Researchers then compared the exposure with death statistics collected by the Institute for Health Metrics and Evaluation. The analysis estimated exposure to DEHP contributed to 350,000-plus global deaths in 2018 among men and women in the 55 to 64 age group. Lead study author Sara Hyman writes, *"By highlighting the connection between phthalates as a leading cause of global deaths, our findings add to the vast body of evidence that these chemicals present a tremendous danger to human health."* (LaMotte, 2025).

Very little has been done to limit the presence of phthalates in consumer products. One type of phthalate was dis-allowed in the production of plastic toys (rubber ducky), in which infants' hand to mouth nature put them at risk. No one is left untouched by their presence. Exposure to phthalates and EDCs is problematic for men. Dr. Shanna Swan has linked exposure with decreasing human fertility levels due to falling sperm counts, decreased sperm quality, and sperm with damaged DNA (Swan, 2021).

Because of the ability for phthalates to cross the placental barrier, it is contingent upon women of child-bearing years to limit their exposure. Realistically, this means women must become cognizant of phthalate-laced consumer products. Women use the preponderance of phthalate-laced products. For example, adult women use approximately 12 individual personal care products daily, while the average teenage girl uses 17 products. As these girls transition to adult women, the potential exists for them to become fetal incubators (Feldscher. 2016). Just as women are warned against the use of alcohol, drugs, or smoking during pregnancy, they must take particular care with their purchase and consumption of processed foods, time-release medications, cleaning products, and most importantly, their use of personal care products, cosmetics, beauty care products, and all scented or fragranced products. It presents quite a list of pregnancy don'ts. Dr. Nathaniel DeNicola, American College of Obstetricians and Gynecologists, writes, *"The toxic burden on pregnant women is a clear and immediate health-care concern. Fetal development is a critical window of human development, and so any toxic exposure during pregnancy does not only have a short-term effect at that moment, but really an effect that lasts the entire lifetime"* (DeNicola, 2018).

Phthalates in our lives exist at the tip top of the EDC iceberg. It is estimated 1,000-plus of the synthetic chemicals produced act as endocrine disruptors (Endocrine Disrupting Chemicals in the Home, 2021). For women of reproductive age, the prenatal (*in utero*) environment must be pristine (no alcohol, no smoking-derivatives, no caffeine, no drugs, and no EDCs). In other words, what we as women reap (*in utero*), we sow for our babies later in life. In addition, men, you do not receive a get out of jail free card. When exposed to phthalates and other EDCs, reduced sperm counts, poorly swimming sperm, and deformed sperm are reducing human fecundity or reproductive potential. Toxic exposure is narrowing life's wick at both ends of life's candle. Indeed, this has ramifications for the human species.

Playing with Hormones... Preys on Metabolism

Worldwide obesity has more than doubled since 1980; while in the United States, over 35 percent of adults and 17 percent of our children are obese. Research indicates EDC exposure plays a role, especially exposures to obesogens and diabetogens. Obesogens are chemicals which upon entry into the body disrupt normal lipid metabolism. Diabetogens enter the body and kill pancreatic ß-cells or disrupt their function, contributing to the potential of diabetes. EDCs disrupt and block normal hormone and receptor communication, especially those associated with the thyroid gland. With EDCs in place, instead of native hormones, proper metabolism of lipids, energy balance, and glucose sensitivity become compromised. Thus, without proper maintenance of this system, we are predisposed to gain weight. Termed metabolic syndrome, this increases the likelihood for the development of type 2 diabetes and cardiovascular disease. Furthermore, the exposure is not just a one off, EDC exposure in the womb, can lead to obesity later in life ("Impact of EDCs on Metabolism and Obesity.")

As highfalutin as obesogens and diabetogens sound, bleach or sodium hypochlorite has been a common household disinfectant used for decades. It also is suspected to contribute to endocrine disruption. Chlorine released from bleach and iodine, a crucial element required for proper thyroid function, are both members of the halogen family of chemical elements. It is hypothesized that the body can mistake chlorine for iodine, inhibiting the uptake of iodine (iodide) by the thyroid gland. This compromises the production of thyroid hormones which are required for regulation of the body's metabolism. Additionally, commercial bleach is often contaminated with perchlorate, another recognized thyroid inhibitor or endocrine disruptor (Greiner, 2008).

Because bleach is easily affordable and available, it is widely used for household disinfection. However, do not mistakenly confuse disinfection with cleaning. For bleach to be effective as an anti-pathogen or disinfectant, it must be used on surfaces that have been cleaned with the removal of dirt and grime. Besides its suspected role in endocrine disruption, bleach acts as an asthmagen. In 2012, the Association of Occupational and Environmental Clinics (AOEC), determined bleach not only acts as a trigger to induce an asthma attack; inhalation of bleach over time develops asthma. Bleach also acts as an irritant to skin and the eyes. From personal experience, my asthmatic daughter often developed asthmatic episodes after swimming in chlorinated pool water. Because of my daughter's suspected asthmatic trigger, bleach was not used in my home.

Children (5 years and younger) are at greater risk from breathing bleach fumes because of the immature development of their lungs. To illustrate, exposure to bleach was the cause of 31,224 calls to United States Poison Control Centers in 2019. One third of these calls occurred for children under the age of five. Bleach, by itself, is toxic; however, when it is mixed with other common household chemicals it becomes extremely toxic, perhaps fatal. Mixing bleach with ammonia, quaternary ammonia compounds (found in other disinfectants), household vinegar or other acids, can create toxic fumes which act directly as respiratory irritants when inhaled.

Bleach is corrosive to a myriad of surfaces including many types of metals. Therefore, bleach should never be used on stainless steel, aluminum, copper, or brass surfaces. It also corrodes marble and granite. Dirt and grime laced surfaces neutralize the effects of bleach, negating its use as a disinfectant. It bears repeating, bleach is not a household cleaner. Additionally, because of its chemically limited shelf life, upon purchase, it must be used within a 3-month timeframe ("Green Cleaning, Sanitizing, and Disinfecting Fact Sheet").

Oxybenzone, is an environmental contaminant found in sunscreens and personal care products, used to minimize the damaging effects of ultraviolet (UV) radiation. Its prevalent use in sunscreens, not only has negative implications for human health, but presents negative effects for inhabitants of aquatic ecosystems. Oxybenzone can react with chlorine, producing toxic by-products in swimming pools and wastewater treatment facilities. Once rinsed from the bodies of human bathers and swimmers, it becomes a contaminant of fresh and saltwater communities. For example, oxybenzone increases mortality of coral organisms, resulting in the phenomena of coral bleaching. Additionally, the contaminant can revisit humans via the food chain with the consumption of oxybenzone-contaminated fish.

Concerning human health, oxybenzone is associated with contact and photo-contact skin allergies. It has been linked to Hirschsprung's Disease, and it is a suspected endocrine disruptor. Hirschsprung's disease causes chronic constipation and abdominal distension in infants. Severe cases cause nausea, vomiting, and lead to poor weight gain. It becomes life threatening, if the intestine presents a bacterial infection. The potential for Hirschsprung's disease is a very good reason to never expose a baby or infant to an oxybenzone product.

With the production and addition of micronized zinc oxide and titanium dioxide as active anti-UV ingredients in sunscreens and other personal care products, oxybenzone's use has rapidly decreased (DiNardo, 2017). Hawaii banned oxybenzone in sunscreens in 2021, after studies found oxybenzone can bleach and deform coral reefs (McConnell, 2023).

Synthetic parabens were developed as cheap preservatives to prevent lipid-based consumer products from becoming rancid. Because parabens act as biocides to inhibit fungi and bacteria, they exist in a plethora of products for the extension of product shelf life. Parabens are used in a variety of 'apply and rinse off products' like shampoos and conditioners. Additionally, they are added to cosmetics, moisturizers, face cleansers, creams, lotions, sunscreens, deodorants, shaving gels, toothpastes, pharmaceuticals, and even some food products (beer, soft drinks, sauces, jams, pickles, and frozen dairy products).

Once we apply paraben-laced products, they readily dissolve and are absorbed across the skin barrier to enter the body. In a study of women with breast cancer who used underarm deodorants containing parabens, 158 of the 160 breast tissue samples analyzed found 99 percent of the samples tested positive for at least one paraben ester. Parabens demonstrate "estrogen agonist" properties, or the ability to commandeer and bind to estrogen receptors in cells, such as breast tissue cells. It is suspected the binding of parabens to estrogen receptors may lead to cell proliferation and growth in breast cancer tissues (Mandai, 2012).

Parabens also possess androgen antagonist activity which has reproductive ramifications. Researchers collected sperm and urine samples from 325 men who attended a Polish infertility clinic. Sperm samples were analyzed for quality and condition of sperm, while urine samples were analyzed for the presence of paraben derivatives. Men presenting the highest levels of urine paraben derivatives were more likely to have abnormally-shaped sperm and sperm having questionable swimming ability. Both criteria reduce reproductive potential. Additionally, urine samples with the highest paraben content, especially levels of butylparaben, were associated with sperm having damaged DNA. Acting as endocrine disruptors, parabens may trigger inappropriate hormone signaling, thus affecting testosterone levels and sperm production. Even though metabolized by the body, their ubiquitous presence in so many consumer products present as another pervasive/everywhere toxic exposure (Stoiber, 2019).

Bisphenol A or BPA is an industrial chemical formulated in the 1950's. It is found in polycarbonate plastics and it is used as an epoxy resin. These plastics found their way into containers that store foods and beverages. Thus, their prevalent use in plastic water bottles. Epoxy resins are used as coating inside metal products, such as food cans or the lining of can tops. Some dental sealants and composites may also contain BPA. Our exposure to BPA comes in its ability to leach from plastics. Once it enters the body through ingestion, it is suspected to disrupt thyroid function, linking it to development of metabolic syndrome, type 2 diabetes, and cardiovascular disease. BPA's ability to cross the placenta is a concern because of possible health effects on the brain and prostate gland of the fetus. Its toxic presence warranted development of BPA-free plastics, especially in water bottles and other products ("What is BPA, and what are the concerns about BPA?" 2023).

Just as phthalates serve as the poster child for everywhere chemicals, PFAS serve as the present-day poster child for persistent/forever chemicals. PFOA or perfluorooctanoic acid was invented by the 3M company and then sold to DuPont Chemicals in the 1950's. Perfluorooctanoic acid is commonly referred to as C8 and it is used in the manufacture of Teflon, fire-fighting foams, lubricants, waterproof fabrics, stain-resistant textiles, carpeting and drapery textiles, and food packaging materials.

Both 3M and DuPont recognized C8 presented a disposal problem. However, for decades, DuPont pumped hundreds of thousands of pounds of PFOA powder into the Ohio River where the chemical found its way into animals (domesticated and wild), the environment, and unfortunately, people. Residents of Parkersburg, WV, adjacent to the Ohio River, were dying of rare cancers, suffering birth defects, developing blackened teeth, and suffering other health maladies for decades due to the toxic forever presence of PFOA compounds in their water supply. After testing 70,000 people from the Parkersburg community, researchers concluded a probable link existed between PFOA contamination and an increased risk of kidney and testicular cancers, thyroid disease, hyper-cholesterol levels, pre-eclampsia, and ulcerative colitis (Wambolt, 2021).

Class action lawsuits brought against PFAS manufacturers 3M, DuPont, Chemours, Corteva, and BASF have resulted in payouts numbering in the billions of dollars for those communities and individuals adversely impacted. Significant sums have been allocated to public water systems for cleanup and remediation, while individuals received compensation based on injurious health effects (Hardin, 2025).

This Photo by Unknown Author is licensed under CC BY-NC-ND

Although we know the presence of PFAS compounds in

our drinking water is harmful, federal protections continue to side with public water utilities to slow their removal; specifically, the removal of four of the forever chemical PFAS-types. According to Betsy Southerland, a retired official from the EPA's water office, "*We're just going to have to keep drinking PFAS for another decade*." Melanie Benesh, a representative of the Environmental Working Group, notes 158 million Americans drink water contaminated with forever chemicals (Fast, 2025).

Even the city of Bemidji, MN where the author resides, found its 5-well water supply contaminated by PFAS chemicals. The city's wells, located adjacent to the airport, were contaminated with PFAS from firefighting foam which had been applied by firefighters for decades. Use of the foam was discontinued in 2009; however, the 'forever' contamination was left behind in the city's water supply (Marohn, 2019). Ultimately, the city built a state-of-the-art water treatment plant for the successful removal of PFAS with funds procured from 3M after prolonged legal negotiations (Albrecht, 2022).

PCBs first manufactured in 1929, were used in a wide range of industrial processes. Because they were non-flammable, chemically stable, and had a high boiling point, they demonstrated exceptional electrical insulating properties; thus, their presence in electrical, heat transfer, and hydraulic equipment. However, their chemical stability signaled their inability to degrade easily. Consequently, it is appropriate to designate PCBs among the first problematic persistent/forever chemicals. PCBs found their way directly into homes as plasticizer agents in paints and plastics, carbonless copy paper, dyes, and other pigment-laced products.

PCBs were prevalent as chemical synthetics for fifty years, until their ban by the EPA in 1979. Consequently, these persistent chemicals remain today in industrial materials and machinery manufactured before 1979. They remain toxically present in sediments of waste sites and bound in aquatic sediments. Therefore because of bioaccumulation, the potential for its toxic presence occurs with the consumption of contaminated fish. Old transformers, capacitors, voltage regulators, switches, appliances, fluorescent light ballasts, cable insulation, thermal insulation materials, adhesives, oil-based paints, caulking, plastics, carbon paper, and floor finishes are all sources of PCBs. For the EPA to ban its production, exposure to PCBs must have demonstrated a wide-range of harmful health effects. These included: endocrine disruption (EDC), decreased immunity, reduced birth weight in babies, reduced sperm counts in men, neurological learning deficits, and promotion of cancer (carcinogen) (epa.gov/pcbs, 2025). My advice to consumers, beware of purchasing a home built prior to 1979, due to the most likely presence of PCBs (Markowitz, 2018). Please carefully consider what forever really means.

Table 1: Endocrine Disrupting Compounds (EDCs)

Synthetic Chemical	Common Household Exposures	Health Effects
Phthalates (Pervasive)	Fragranced Products Vinyl Products Plastics	Negative effects upon reproduction, fetal neurodevelopment, and cardiovascular issues in adults
Parabens (Pervasive)	Personal care and beauty products Cosmetics Foods	Demonstrates estrogen agonist properties and androgen antagonist properties
PBDE (polybrominated diphenyl ethers) (Persistent)	Fire retardant in foaming materials in furniture, carpet, bedding, and draperies Use being phased out.	Neurodevelopmental issues
Atrazine (Persistent)	Herbicide in corn and sugarcane production Groundwater and surface water contaminant	Interferes with hormonal systems in animals, resulting in reproductive anomalies
BPA (bisphenol-A) **Short half-life, widespread use**	Lid and can liners Store receipts	Fetal neurodevelopmental effects Increased incidence of hypertension, Type 2 diabetes
PCBs (Polychlorinated diphenyl) **(Persistent)**	Anything electrical prior to 1979, oil-based paints and caulking prior to 1979 Contaminated fish	Carcinogen Birth Defects Reduced fertility Immunity compromised
PFAS (per- and polyfluoroalkyl) **(Persistent)**	Drinking water Water/stain resistant fabrics, Teflon products	Thyroid disease (metabolic syndrome, Type 2 diabetes, cardiovascular disease) Cancer (kidney, testicular)
Oxybenzone **(Pervasive)**	Sunscreen Cosmetics	Contact and photo-contact skin dermatitis Suspected EDC
Sodium hypochlorite, Bleach	Household Disinfectant Bleaching Agent	Thyroid disruption Asthmagen

(Pervasive)	Swimming Pool Sanitation	Respiratory, Eye, Skin Irritant

Other Household Toxic Exposures

Consumers should be aware of other red flag toxins that are hidden in common household products, especially our personal care and beauty/cosmetic products. Besides EDCs, a list of other red flag synthetic toxins in and around the home must include: the chemical preservative formaldehyde, diethanolamine, heavy metals (mercury, arsenic, cadmium, aluminum, and lead, etc.), coal tar found in synthetic dyes, and sulfate surfactants, such as sodium lauryl sulfate (SLS) and sodium laureth sulfate (SLES) used in cleaning, personal care products, and cosmetics.

The embalming preservative formaldehyde is found in many cosmetic and personal care products. In 2011, the United States Department of Health and Human Services declared formaldehyde a human carcinogen suspected to cause nasal and nasal pharyngeal cancers. This chemical serves as a preservative in personal care products and cosmetics. Formaldehyde and formaldehyde-releasing preservatives (FRPs) are found in nail polish, body washes, hair conditioners, shampoos, skin cleansers, eye shadows, and many other personal care formulations ("Toxicological Profile for Formaldehyde"). It is estimated that formaldehyde as a cosmetic preservative is found in 20 percent of all cosmetics produced in the United States. In addition, on average, the American female uses 16 to 20 different personal care products each day, including cosmetics (Murray, 2017). When do we introduce such products to our younger females?

They say that "beauty" resides in the eye of the beholder. This means that our perception of beauty is subjective; we can differ in its definition. Historically, beauty has been defined in ways that are intertwined with social and cultural mores. Ancient civilizations valued beauty beyond physical attributes to include intelligence and demonstrating virtuous, moral character. Our modern-day standards for beauty are not so virtuous, as they are influenced by social media and cultural messaging. This has skewed our definition of beauty in a most unfortunate and toxic way, especially with our children and teenagers. This definition has forced us to cover up and mask what we perceive as physical inadequacies or imperfections. To illustrate, 'gray is beautiful,' is not a beauty standard embraced by women as we age. Instead, women are willing to apply hair dyes that often contain a neurotoxin, such as lead acetate, and please consider where that neurotoxin is being applied.

Licking one's lips is akin to blinking one's eyes. We do both often throughout the day as involuntary reflex actions. Many lipstick formulations are laden with heavy metals including: lead, cadmium, chromium, manganese, and aluminum. The metals are bound within the mineral dyes used to create the fantastic, diverse array of lipstick colors. An FDA study conducted on 400 different lipsticks found the presence of neurotoxic lead compounds (fda.gov/cosmetics, 2022). The same toxic stuff used to hide the gray. The same toxic stuff that is so toxic, that its use was banned in gasoline and house paints long ago. Yet, there it is in a cosmetic that is applied and re-applied as we continually lick it off.

Yes, beauty resides in the eye of the beholder. On average, American women spend $3,756 annually to achieve their individual standard for beauty. It is most unfortunate that women, as consumers, do not attach a value for what personal care and cosmetic products are doing to their insides. This mis-prioritization comes at a truly higher, incalculable cost for women's health and wellbeing and most unfortunately, that of a developing fetus.

Lead and other heavy metals such as, arsenic, cadmium, mercury, chromium, zinc, and iron are found in a wide variety of consumer products, including personal care, and beauty products. In a survey conducted by a skin care retailer, it is estimated American women, on average, convert their $3,756 annual expenditure into a staggering $200,000 (the cost of a house) expenditure in a lifetime, just for the use of daily-applied personal care and cosmetic products. On average that is 16 products a day with an exposure of 100 different chemical synthetics each day (Murray, 2017). Facial beauty products include: lip products, eyeliner, nail polish, foundations, eye shadows, blush, concealer, and moisturizers. One recent study, focusing on girls only, had them hand over their lipsticks and glosses for heavy metal analysis. Metals were present in varying degrees depending on the cosmetic brand. The study indicated for those girls/women who apply lipstick and glosses often (3 or more times/day), they surpassed the daily recommended exposure for chromium, aluminum, and manganese.

Most concerning was the presence of lead in 75 percent of the samples tested (Liu, 2013). Lead, as a neurotoxin, is extremely injurious to infants and children. Its neurotoxicity has been linked to learning, language, and behavioral deficits in children. This is especially relevant for children living in older homes where lead-based paints and lead pipe water systems still exist. On a community level, Flint, Michigan stands as an unfortunate example we all must heed. Additionally, if lead has not been remediated, it resides in household dust impacting the home's inhabitants. Lead has been linked to reduced fertility in both men and women and pregnant women are especially vulnerable, since lead can cross the placental barrier.

In personal care and beauty products, heavy metals may be added as ingredients or they simply exist as contaminants. Metals, as contaminants, have been found in products including: sunscreen, foundation, nail polish, lipstick, and whitening toothpaste. Different brands of commercial toothpastes and powders recently were analyzed for the presence of heavy metals. Of the 51 brands analyzed, 90 percent tested positive for lead, 65 percent tested positive for arsenic, 47 percent tested positive for mercury, and 35 percent tested positive for the presence of cadmium. Thank goodness, we teach our children to brush, rinse, and spit (Bassi, 2025). In addition, metals, such as chromium, aluminum, and iron serve as colorants in eye shadows, lip glosses, lipsticks, and nail polishes.

Even though we require some metals for proper function, blood oxygenation, and iron, for example; it is the unwarranted exposure and absorption of heavy metals that is problematic. To illustrate, biopsies of cancerous breast cells show higher accumulations of heavy metals, such as iron, cadmium, zinc, and mercury. Mercury, found in some skin lightening creams, is not only a neurotoxin, it is toxic to the reproductive, immune, respiratory, and endocrine systems. Because mercury is readily absorbed across the skin, its use in cosmetics is concerning. It is very problematic for pregnant women, since it may cross the placental barrier presenting negative effects on fetal development. Aluminum-based compounds are linked to neurotoxicity, cancer, and developmental and reproductive problems. Chromium is strongly linked to immune and respiratory toxicity. Besides, alcohol-use and smoking, let us add that women who are contemplating pregnancy should avoid not just EDCs in the home, but the presence of heavy metal compounds, especially those found in personal care products (safecosmetics.org).

Diethanolamine (DEA) is a common ingredient added to personal care products, such as sunscreens. Its action as a foaming agent explains its presence in bar soaps, foaming soaps, as well as shampoos, cleaners, detergents, polishers, cosmetics, and pharmaceuticals. Once exposed, we absorb DEA across the skin and the scalp. Acute health effects to diethanolamine exposure include: irritation of the eyes, burning of the skin, irritation to the nose and throat, resulting in coughing and wheezing response. Because soaps and shampoos are lathered onto the skin, diethanolamine may result in skin allergy presenting as an itchy rash. Furthermore, diethanolamine exposure may lead to headaches, nausea, and vomiting. Long term health effects in humans have not been well documented, but animal testing supports damage to the male reproductive system, liver, blood, and the central nervous system. The National Toxicology Program (NTP) reported increased incidence of liver and kidney tumors in mice from dermal exposures to diethanolamine. However, the EPA has not classified DEA as a carcinogen (NJ.gov/health/eoh/rtkweb/documents/fs/0686.pdf).

Another toxic category to consider is the presence of synthetic dyes in the home. Color additives include dyes, pigments, or other substances that impart color to a food, drug, cosmetic, and consumer-based products. Color additives make products more appealing to the eye. As consumers of food, we eat with our eyes before we ever sit down at the dinner table. Even more compelling, as consumers of products, we also purchase with our eyes before we ever evaluate the product's potential harm.

Historically speaking, natural dyes occurred with substances such as turmeric and paprika, while mineral sources included the use of iron and lead oxides. By the 1800s, organic dyes used in foods, drugs, and cosmetics were produced from by-products of coal processing, referred to as coal-tar colors. Some colorants were added to hide imperfections, such as their additions to mask inferior food products. By the 1900s, many foods, drugs, and cosmetics were being color-enhanced with the use of synthetic colorants. Akin to today's greenwashing, this is yesteryear's color-washing. To illustrate, margarine, an artificial replacement for butter, was formulated in vivid yellow compared to butter's more pallor palette. Besides, their presence in foods, which is concerning enough, synthetic colorants exposed the populace to lead, arsenic, chromium, mercury, and other metals in the creation of hair dyes, foundation, rouge, and eye cosmetics (Barrows, 2003).

Coal-tar derivatives, as an additive in any consumer product, should make each of us take pause. It has been added to consumer products since the 1800s. Everyday products, especially cosmetics, contain coal-tar derivatives, identified by "CI" in the list of ingredients. Synthetic coal-tar dyes, now produced from petroleum, once were produced by combining aromatic hydrocarbons, like toluene, xylene, and benzene, which are extracted from the distillation of bituminous coal. Coal-tar, by itself, is a recognized carcinogen. Petroleum coal tar dyes remain an ingredient in a plethora of consumer products, primarily hair dyes and cosmetics. Coal-tar dyes color our fabrics, makeup and hair, lipsticks and glosses, and paints. (Suzuki, 2025).

On April 22, 2025, HHS (Department of Health and Human Services) and the FDA (Food and Drug Administration) announced a series of efforts to remove all petroleum-based synthetic dyes from the American food system and replace them with natural alternatives. Three dyes including: red dye-40, yellow dye-5, and yellow dye-6, are presently used in 90 percent of American processed foods and they are being phased out. Synthetic dyes, particularly harmful to children, are suspected in food allergies and in the promotion of developmental issues, such as ADHD. Consequently, the FDA is fast-tracking approval of natural dyes such as, calcium phosphate (a whitening agent), Galdieria extract blue (from algae), Gardenia blue (plant extract), and Butterfly pea flower extract (already in use as a colorant in soft drinks, ice cream, candy, and yogurts). These changes are long overdue and the American consumer will have to adapt to non-vibrance in America's processed foods. To illustrate, Cheetos no longer will radiate flaming orange hues, while the lustrous red of maraschino cherries will pale to pink (Ducharme, 2025).

Sulfate surfactants, Sodium lauryl sulfate (SLS) and its milder sister chemical, Sodium laureth sulfate (SLES) are found in more than 90 percent of personal care and household cleaning products. Surfactants are chemical agents that help substances, having different chemical characteristics, mix and spread more easily. If you see foams with the use of soap, shampoos, body washes, bubble bath, and detergents, then SLS or SLES are probably present. They also are present in acne cleansers, shaving cream, lip balm, hand sanitizer, makeup remover, foundation, facial cleansers, exfoliants, toothpaste, hand cream, sunscreen, even some mascara products. Absorbable across the skin, SLS and SLES may act as skin irritants. Skin irritation is related to the amount of time they persist on the skin, as tests have shown prolonged presence on the skin is more likely to induce mild/moderate skin irritation. Consequently, the application of SLS or SLES is of less concern in 'apply and rinse' products, such as shampoos, soaps, and body washes. (Schaefer, 2023). Because SLS and SLES can be produced from coconut oil, it encourages manufacturers to disguise SLS or SLES listings, simply as coconut-derived ingredients. Do not be fooled. The chemical processing producing SLS or SLES results in surfactants much harsher to the skin than coconut oil.

There you have it, a far-ranging, diverse list of toxic chemicals found in products lining our household shelves and even present in the refrigerators of every American home. By the time we reach adulthood, we have applied 100's of quarts of synthetic-based lotions, shampoos, cosmetics, and other personal care products onto our skin. Females expose themselves to over 100 different synthetics each day, simply through their use of personal care and beauty-based products. Add in household cleaning products and a truer, more realistic picture comes into focus. Our reality, the household air we breathe, the dust we ingest, the products we apply, the surfaces we clean, and even the foods we eat contain chemical synthetic toxins. Toxically-speaking, what we find residing inside of us or alternatively, in the waste stream leaving our homes derives as a by-product of living in a modern, industrialized, consumer-based society (Smith, 2025).

Each chemical synthetic, by itself, can present a toxic presence within our body. However, what is the outcome of the continual collective daily exposures from our home's synthetic chemical cache and what about their synergistic actions together? This is the focus of the SELMA study conducted in Sweden and other European countries. Data collected from SELMA provides evidence of neurodevelopmental issues with delays in speech and behavioral issues occurring in babies from mothers exposed to a chemical cocktail of EDCs. Troubling results of the study imply synthetic toxins collectively and synergistically compound epigenetic modifications in the genes of the mother, as well as fetal genes. This emphasizes the need for increased attention to lifestyle choices (beyond no smoking, no drugs, and no alcohol) which must be communicated by those who serve expectant mothers and their precious cargo (Engdahl, 2021).

After the fact, the price we pay for us and our children is incalculable in terms of our health and wellbeing. It does little good for us to read and process this information and then do nothing. Instead, as the Rolling Stones remind us, "*time waits for no one*." By internalizing this information, you have taken an important first step by becoming a better informed and aware consumer. Now, you must pay it forward and take the step that reduces you and your family's exposure to synthetic chemical toxins. Evaluate your home's chemical toxins, prioritize the toxic products for elimination, and then decide **what swap will you make** in your home today?

Table 2: Additional Toxins of Concern

Many of these agents are found in a multitude of household products. They are added as preservatives to increase product shelf life. They act as colorants or provide chemical character, such as foaming action. Acute effects often include irritants of skin, eyes, nose, and throat leading to rashes, headaches, nausea. Most serious chronic effects are neurotoxic and carcinogenic in nature.

SYNTHETIC CHEMICAL	COMMOM HOUSEHOLD EXPOSURES	HEALTH EFFECTS
Diethanolamine (DEA)	**Foaming agent** found in sunscreens, foaming soap products, cleaners, detergents, cosmetics, pharmaceuticals	**Acute effects:** Eye, Nose, Throat Irritant, Skin allergy, headaches, nausea **Chronic effect:** Animal studies support reproductive, nervous, hepatic, and circulatory dysfunction
Heavy Metals • **Lead**	May be added as ingredients or exist as contaminants in a diversity of cosmetics, beauty,	**Lead is a neurotoxin.** **Mercury is a neurotoxin.**

• **Mercury** • **Chromium** • **Cadmium** • **Arsenic** • **Zinc** • **Manganese** • **Iron** • **Aluminum**	and personal care products. **Colorants** used in moisturizer, deodorant, toothpaste, nail polish, lipsticks, glosses, eye shadow and liner, mascara, foundation, blush, concealer	**Accumulated heavy metals** in body tissues are linked to reproductive, immunity, and respiratory issues. Possibly carcinogenic
Coal Tar Synthetics	**Dyes** in foods and cosmetics **Colorants** added to foods, textiles, make-up, hair dyes, lipsticks and glosses, and paint	Food allergies Learning and developmental issues (ADHD)
Sulfates **Sodium Lauryl Sulfate (SLS)** **Sodium Laureth Sulfate (SLES)**	**Surfactants** found in more than 90 percent of personal care and household cleaning products. Foaming action of soaps. Also found in a diversity of cosmetics.	Skin irritants, especially in products that are not applied and then rinsed from the skin.
Formaldehyde Preservative	Pressed wood products Tobacco products, smoke **Preservative** in personal care products: body wash, conditioner, shampoo, skin cleanser **Preservative** in cosmetics: nail polish, eye shadow	**Acute effects**: eye, nose, throat irritant, headaches, dizziness, nausea **Chronic effects**: carcinogenic (nasal and pharyngeal cancer agent)

America's Pill Culture

Commercial household products have been scented to mask the real scent of their not-so-pleasant chemical formulations. Ammonia or bleach presenting a floral note or a hint of pine, appear to make them less noxious. However, these more pleasant scents promote their use and our body's response is one of distress. Many have experienced some type of acute distress induced by scented products, such as watering of the eyes, tightening of the throat, precipitating a sneeze or a cough. Those signals were alerting us to take care and be aware. However, with continual, repeated exposures, the response is heightened to include: headaches, breathing difficulties, nausea, or development of skin rashes.

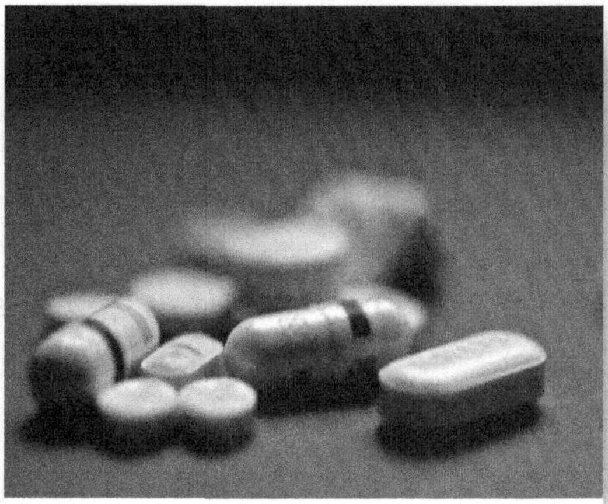

Instead of disengaging ourselves from the offending stimulus, our response is to treat the symptoms, while continuing the exposure. Stimulus and response are at the very core of survival; yet, our response is to treat the symptoms. Thus, begins entry into a very precarious and vicious cycle. OTC (over-the-counter) medications made an estimated revenue of nearly 41 billion dollars in 2024. They are America's indispensable allies to counter the daily sniffles, a scratching throat, or a little twinge of the back. We often purchase OTC meds to address symptoms of the cold or flu, allergies, heartburn, sinus pressure, constipation, and to provide pain relief (Bansal, 2024).

OTC meds are one component of America's pill problem. Convenience, grocery, and of course, drug stores have dedicated aisles for our review. In fact, Americans make 2.9 billion trips annually for OTC purchases, at a cost of $338.00 per family. We buy them, we use them, and while not in use, they occupy a household cabinet or shelf in bottles containing 100s of them ("Five Consumer-Facing Challenges of the OTC Medicine Experience." 2019). Yes, America has a pill problem and abuse it; we do. It is interesting to note that a medical questionnaire asks us how often we drink an alcoholic beverage, smoke tobacco products, but not **how often** we reach for a bottle of acetaminophen, ibuprofen, laxatives, or TUMS for the tummy. Dependency is as dependency does.

There are five categories of OTC medicines: analgesics, cough medicines, sedatives, antihistamines and decongestants, and laxatives. Excessive use of analgesics such as acetaminophen or ibuprofen may result in liver damage (acetaminophen) and kidney damage (ibuprofen). Over use of laxatives can lead to electrolyte imbalances and dehydration. This presents a secondary gastrointestinal upset, then induces an over-correction and one pill leads to another. Unfortunately, dependency on OTC medications can lead us down the path to more addictive medications (Cooper, 2011).

America, by far, has the dubious distinction of leading the world in taking opioids. Do American doctors over-prescribe their use? Consider the case of Kermit, West Virginia. According to a congressional committee, in two years, the 400 residents living in Kermit received almost nine million opioid pills. This example and the wide availability of pills, OTC or prescribed, point to a failed health care system. While private insurance coverage may not cover a visit to a physical therapist, a chiropractor, an acupuncturist, or another alternative treatment for a patient presenting a muscle twinge; insurance companies will and do pay for pills. The United States constitutes roughly 4.4 percent of the global population, while it consumes over 80 percent of the world's opioids. In 2015, Americans were responsible for 99.7 percent of the world's hydrocodone consumption. The statistics are a staggering condemnation of a health care system gone awry.

The United States and New Zealand are the only countries that allow prescription drugs to be advertised on television. Pharmaceutical companies spend billions annually to advertise drugs to counter issues from dermatological skin issues to constipation to allergies, and the list goes on. The result, we self-diagnose and then ask a clinician for the pill advertised on the television. An exceptional example of advertising's role in America's pill problem was provided with a 60-second ad during the 2015 Superbowl. AstraZeneca had an answer, in pill form, to address opioid pill-induced constipation. The ad advised sufferers to visit their doctor to ask for a pill to address an issue caused by another pill. Pill taking is a vicious cycle and it is a cycle we must break (Amos, 2017).

We have a problem not only with the drug inside the pill, but what covers the pill. Phthalates rear their ugly head again, since they are applied as excipients in time-release capsules and enteric pills. Time-release medications use coatings to prevent premature dissolution in the stomach and promote release and absorption through the lining of the small intestine. Phthalates promote flexibility and durability of the exterior of these medications allowing them to circumvent the stomach's acidic environment. Additionally, enteric coatings are widely used with OTC medications known to cause gastrointestinal upset, such as aspirin or naproxen (Aleve). For those individuals who require daily medications for chronic health disorders and for those who take OTC medications for acute disorders, phthalate-laced coatings in time-release medications and in OTC enteric coating medications present another degree or degrees of household toxic exposure (Geya Sree, 2023).

This topic is important and relevant because our reliance on pills presents another form of synthetic chemical contamination. Drugs, OTC or prescription, pass through us, just like endocrine-disrupting compounds (EDCs) becoming part of the waste stream. It should not be surprising that drinking water is tainted with hormones from birth control pills, antibiotics, antidepressants, immunosuppressants, hypertension drugs, acetaminophen, and others. To illustrate, a recent modeling study forecasts that each year, the earth's waterways (fresh and salt) will receive thousands of tons of antibiotics. In fact, 8,500 tons are projected to leach into the world's river systems (Price, 2025). Antibiotic contaminants leaching from farmers' fields, un-used pills thrown away in the garbage, nursing home, and hospital waste, and tainted human urine, all contribute antibiotics and other drugs into the waste stream. Add PFAS and other forever chemical contaminants to the waste stream and we must question, at what cost to aquatic species and to the 'fresh' water we drink?

The results of the potential long-term exposure to small increments of pharmaceuticals remains unknown; yet, it is valid, concerning, and worrisome. America's 170,000 public water systems are monitored for nearly 80 harmful substances including: bacteria, viruses, pesticides, petroleum derivatives, and some metals. What about fragrance derivatives, lotions and sunscreens, other personal care products rinsed from the body, and what about drugs?

Sewage treatment plants are not designed to remove pharmaceuticals from water, nor are the plants that treat water to make it potable. However, some research indicates that conventional treatment methods result in a 90 percent decrease in the amount of ibuprofen and naproxen in the water discharged from wastewater treatment facilities. On the other hand, other drugs such as the pain reliever, diclofenac, are not removed. Drinking water treatment utilizes chlorine to kill bacteria and other pathogens and its presence seems to degrade acetaminophen, codeine, and some antibiotics. However, a 2010 report from scientists working for the Southern Nevada Water Authority found trace amounts of antidepressants, antipsychotics, antibiotics, beta blockers, and tranquilizers in drinking water after treatment ("Drugs in the Water." 2011).

Fragrance derivatives, lotion and sunscreen derivatives, OTC meds, anti-this and anti-that are found in our drinking water. Is there a cumulative effect from this witches' brew? We do not know, but recall mothers and fathers mix baby formula with water and feed it to their babies. America has a pill problem. It provides yet another rationale for us to clean up our act and choose an alternative path for our day-to-day living.

Be Realistic in Prioritization

Being cognizant of marketing tactics, such as greenwashing, should preclude and prevent a "throw caution to the wind" mindset when contemplating change in our consumer buying habits. Our approach must be prudent with the adoption of a realistic, strategic plan. Laws and regulations, (federal, state, and local) do prevent the presence of some recognized toxins, such as PCBs, lead, and even some food dye exposures. However, is it realistic for us to live our lives without the presence of toxins found in our everyday consumer products?

As a consumer informed about household toxins, what steps can I implement to reduce my probable exposure on a household-level? For example, armed with knowledge of a toxin's probable presence, I can choose solid wood furniture over pressed wood products. I can search for products having low-VOC (volatile organic compounds). I can investigate products advertised as EDC-free (endocrine disrupting compounds), as a standard for purchase. I can employ an air-purification unit with activated carbon filters, which are specifically designed to remove toxins residing in the air. I can use essential oil diffusers versus the use of chemical air fresheners or candles. Additionally, if the season allows, I can open the windows to naturally circulate fresh air in and move toxic air out. However, what do I specifically do about the presence of toxins in personal care and household cleaning products?

Greenwashing and the trade secrets loophole alerts us to the unscrupulous nature of business. Bottom line, companies are in the business to make money or they quickly will be out of business. In the past, our criteria for purchase included product convenience, familiarity, desirability, reliability, and economy. Living in our modern industrialized, consumer-based reality, forces our purchasing calculus to add toxin-free. Consequently, we must invest in companies and their products that value sustainability, and integrity in the manufacture of their products. In other words, the company must win our trust by demonstrating their alignment with our own sustainable and ethical values.

An informed consumer must be a smarter consumer. Control over household toxic exposures comes first with a realistic appraisal of what is already present in the home. Then, of utmost importance, is to implement a toxin-free standard with purchasing choices. Aligned with change in purchasing habits is trust that a company corresponds with a synonymous toxin-free standard. This includes the purchase of those daily-use products with which we are most intimate: household cleaning, personal care, cosmetic and beauty products. As consumers, we retain purchasing power as an investment for our health and wellness.

To illustrate, the University of California prepared a 10-step plan on how to reduce EDC exposure in the home. One emphasis was the removal of EDC-laced personal care products. Not surprisingly, changing to EDC-free personal care products produced immediate effects. A study involving 100 teenaged girls, indicated when the girls no longer used EDC-laced personal care products, their urinary levels of EDC derivatives precipitously dropped, as illustrated in the table provided below.

Table 3: Removal of EDC products reduced EDC-levels in teenaged girls

Urinary Analysis **Baseline EDCs**	**Urinary Analysis** **After EDC-free/3-day trial**
Phthalates	Urine levels fell by 27 percent
Parabens	Urine levels fell by 45 percent
Collectively, other EDCs	Urine levels fell by 33 percent

(Harley, 2016)

If our home is to become less toxic and therefore, restorative; we must choose to purchase EDC-free and toxin-free products. As a reminder, the following table provides a quick and dirty primer of red flag toxins found in daily-use products.

Table 4: Red Flag Synthetic Chemicals to Target

Synthetic Toxin	Exposure	Found In?	Health Effects
Phthalates (EDC)	**Pervasive/Everywhere**	1) Air Fresheners 2) Plastic clothes 3) Shower Curtains 4) Personal Care 5) Cosmetics 6) Food Wraps 7) Food Containers 8) Anything Vinyl	**EDC** Neurodevelopmental issues (fetal) Reduced fertility Pre-mature mortality

Parabens (EDC)	**Pervasive/Everywhere**	1) Cosmetics 2) Moisturizers, creams, lotions 3) Shaving products 4) Deodorants 5) Hair products	**EDC** Endocrine/hormone disruption, suspected to cause hormonal dysfunction/imbalance.
PFAS, per- and Polyfluoroalkyl substances	**Persistent/Forever**	1) Food packaging 2) Teflon 3) Stain- and water-resistant fabrics 4) Microwave bags 5) Coatings 6) Contaminated water and fish	**Cancer** (Kidney and Testicular) Thyroid disease Birth Defects Preeclampsia
Heavy Metals	**Colorants**	1) Cosmetics 2) Beauty products	Lead & Mercury **(Neurotoxicity)**
Diethanolamine	**Foaming Agent**	1) Cosmetics 2) Foaming products	Eye, nose, throat irritant
SLS and SLES	**Surfactants**	1) Personal Care 2) Cleaning products	Skin irritant

Why Essential Oils?

If we fall back to a 20th century concept or '**less is more**' approach, then we must re-visit our use of plant species in the home. Our ancestors built with timber-hewn wood, ate native plants and plant-based products, and if sick, utilized plant-derived home remedies. Herbal plants provided roots, stems, and leaves, as the basis for drinkable teas and topically applied poultices. The ancient Egyptians applied honey as an antibacterial agent, while using aloe to soothe burns. The scent of thyme was valued for its antiseptic properties and as an aid for sleep. Sprigs of thyme were placed underneath pillows, for example. They also employed thyme as a pain reliever and it was included in many medicinal treatments. Greek scholars placed wreaths of rosemary above their door, as they believed its scent improved memory and cognition. Two of the gifts of the Magi, frankincense, and myrrh, were prized as perfumes, inhaled as incense, and ingested as spices. Indeed, frankincense and myrrh have long been used as anti-inflammatories in many cultures (Bo Cao, 2019).

The Sumerians and Babylonians chewed willow bark for pain relief and to address fevers. Native Americans also valued willow bark as medicine. Willow bark contains salicin, which our body metabolizes into salicylic acid. It is this chemical constituent that induces anti-inflammatory, pain-relieving qualities. Salicin is the precursor constituent to modern-day aspirin, acetylsalicylic acid. Methyl salicylate, commonly known as oil of wintergreen, is produced in plants as an anti-viral defense compound. Another plant extract from the South American cinchona tree was found to be an effective treatment for malaria. The active ingredient, quinine, aided in killing the malarial parasite (National Center for Biotechnology Information, 2025).

Just as we as animals have an immune system, so do plants. Essential oils are plant immune compounds produced by aromatic plant species which act as attractants for pollination, and they serve plants as mechanisms of defense, protection, and healing. Plants are continually threatened by infectious, pathogenic fungi, bacteria, and viruses. However, unlike their mobile animal counterparts, immobile plant species are spatially segregated above and below the ground where pathogenic agents attack from all quarters. Thus, aromatic plants produce and release essential oils to protect from damage inflicted by environmental parameters, as well as biological competitors (Ninkuu, et al. 2021).

Hydrophobic, lipid-based essential oils are chemically recognized for their phenolic and terpene content, two organic compound classes. Phenols, structurally characterized by the presence of a benzene ring, serve the plant in both a defensive and detoxifying capacity. Phenols aid in the cleansing of cell membranes, specifically protein receptors, so that better cell communication is achieved. Essential oils, such as oregano, thyme, and basil are noted for their high phenolic content. Pictured is a benzene ring structure, characteristic of phenolic compounds.

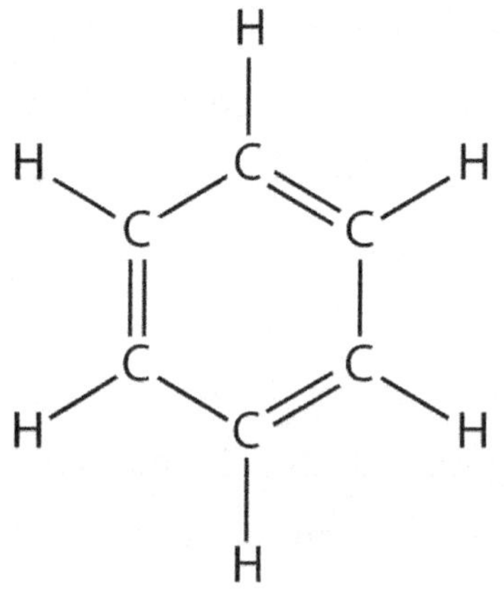

Terpenes, structurally characterized by the presence of one or more isoprene units, contribute to a plant's aroma and flavor. According to the number of isoprene building blocks, terpenes can build from monoterpenes composed of two isoprene units, such as peppermint's menthol, to larger sesquiterpene molecules composed of three isoprene units, such as humulene contributing to the spicy aroma and depth of flavor noted in black pepper. Organic classes, such as phenols and terpenes, promote the building of larger, more complex organic molecules from repetition of these simpler carbon structures.

Monoterpenes and sesquiterpenes dominate the organic structure of essential oils. Whereas phenolic compounds cleanse cell membranes, enhancing cellular communication, terpenes act at the level of nuclear DNA. DNA-corrective terpene molecules act as anti-mutagenic and anti-tumoral agents. Structurally, monoterpenes dominate the design of essential oils; however, their small size encourages rapid vaporization. The pairing and presence of larger sesquiterpene molecules slows vaporization, expanding the attractant and defensive capacity of the essential oil.

Each essential oil is unique in its PMS (phenols, monoterpenes, sesquiterpenes) fingerprint. The PMS fingerprint distinguishes oils from each other in their flavor, odor, and defensive capacity (Hilal, 2024). The beauty of the design is important. Essential oil chemistry is comparable to a musical manuscript. A musical manuscript that is monotonal (one note) would not be very interesting to the human ear; while a manuscript composed of different notes is melodic. Music employs notes of sound, while essential oils employ notes of smell. Extending the musical analogy, think of monoterpenes as high notes in the molecular composition that harmonize with sesquiterpene bass notes.

Phenols in partnership and concert with monoterpene and sesquiterpene molecules make for a powerful and diverse group of immune compounds. Synergy between an oil's component molecules is not additive in effect, but multiplicative. There are approximately 3000 aromatic plants known to produce essential oils (de Sousa, 2023). Because essential oils are simple organic compounds; yet complex; their molecular design promotes their efficacy as molecules of attraction and molecules of defense. This is integral to the survival of the plants that produce them.

The first recorded use of distilled aromatic plant materials or essential oils occurred in medieval Persia (present-day Iran), where the technique of oil distillation was initially developed. Oil distillation was a common-day practice in the Middle Ages during the 11th century. Distilled plant products could be found in European medieval apothecaries. By 1500, oils derived from cedarwood, rose, rosemary, sage, and myrrh were commonly produced pharmaceuticals. Thus, long ago essential oils found a place in pharmacology, a branch of the pharmaceutical sciences, which studies drug action of any synthetic, natural, or endogenous molecule that exerts a biochemical or physiological effect on a cell, tissue, organ, or organism.

The modern-day use of essential oils in aromatherapy evolved to utilize an oil's constituents for pharmacologic, physiological, and psychological purposes bound in an oil's scent character. Aromatherapy first appeared in medical journals in the early 1900's; however, using scents as medicines has been referenced for thousands of years. Egyptians used odors to treat diseases; while the Greek physician, Dioscorides wrote of their healing properties. Essential oils, as molecules of scent, dissolve across the mucus lining of the olfactory epithelium which lines the nasal passageway. The scent molecules stimulate sensory smell receptors which transmit signals to the olfactory bulb. Here, mitral cells transmit output signals to the olfactory cortex, which allows us to perceive and recognize odors. The mitral cells perform an additional function with their direct connection to the brain's center for emotional learning and memory, the amygdala. Once smells become memories, our amygdala in concert with the limbic system, the brain's emotional nervous system, induces emotions which help us to respond and survive in the environment. For example, the smell of baking cookies induces us to smile; while skunk odor causes us to pinch our nose.

Scientific studies do support the use of aromatherapy as an alternative and complementary methodology. A study from Mie University School of Medicine in Japan found that patients suffering depression required smaller doses of anti-depressant medications after inhalation of citrus Yuzu oil. The odor associated with Yuzu's citrus scent reduced heart rate in the patient study group; while having a raised heart rate is a common symptom associated with anxiety (Matsumoto, 2014).

Conifer tree oils, such as essential oils derived from spruce and fir trees, are high in the key ingredient, bornyl acetate. It induces that needle, pine-like odor we so often associate with a freshly cut Christmas tree. In a study conducted by Japanese researchers, individuals performing visual display terminal work (computer screens) were subjected to air having high, low, and no doses of bornyl acetate emitted into the air stream. Bornyl acetate delivery was found to induce autonomic nervous system changes in the subjects. Subjects (workers) were more relaxed without any incidence to task performance even with low dose emission of bornyl acetate (Matsubara, 2011)

A study was conducted to investigate the effects of inhalation of orange essential oil on women's anxiety during labor. In a clinical trial study of 100 pregnant women. The women were assigned to two groups, with one group receiving aromatherapy with inhalation of orange oil, while the other group was exposed to distilled water. Anxiety was assessed with measurements of blood pressure, respiration, and pulse rates. The measurements were taken twenty minutes before treatment to establish a baseline and then measurements were continually taken during and after the inhalation trial. Although the level of anxiety of women in both groups reduced after inhalation intervention, the reduction was greater for the women who inhaled orange essential oil. Researchers concluded aromatherapy is a noninvasive and effective method for women to overcome anxiety during labor (Rashidi-Fakari, 2015).

Cedarwood, characterized by the presence of cedrol, influences the autonomic nervous system. A study performed on women (ages 20s-40s) from three different countries was conducted to monitor this. It appears that cedrol produces a sedative effect in the women studied despite differences in ethnicity and culture. As an index of autonomic nervous activity, the miosis rate (ratio of pupil-diameter variation after light stimulus to initial pupil diameter) in pupillary light reflex was measured before and after cedrol inhalation. The miosis rate significantly increased after cedrol administration, suggesting dominance of the parasympathetic nervous system, inducing the sedative effect (Yada, 2007).

Lavender oil has been recognized for its calming character. Its key constituents are linalool and linalyl acetate. In animal studies, the scent of linalool provides sedative effects aligned with those of diazepam, a drug used to calm symptoms of anxiety. Additionally in a study of hypertensive adults, with inhalation of a blend of the essential oils of lavender, ylang-ylang, marjoram, and neroli, the study population experienced a greater decrease in the levels of the stress hormone, cortisol. With a reduction in cortisol levels, their daytime blood pressure levels normalized (Kim, 2012).

According to the journal, *Current Drug Targets* (2013), olfactory signals from essential oils are thought to impact chemical production in the brain (Nan Lv, 2013). Diffusion of oils, such as pure lavender, is recognized in multiple hospital settings to aid in calming anxious patients. Lavender oil modulates $GABA_A$ receptors in the central nervous system resulting in a quieting or calming of the system. Research published in *Frontiers of Behavioral Neuroscience* indicates diffused pure lavender must be inhaled or smelled to exert calming effects on patients experiencing pre-operative stress or presenting other anxiety-type disorders (Harada, 2018). The following table presents some essential oils providing mental benefits.

Table 5: Benefits of Diffusing Essential Oils

Essential Oil SCENT	RESPONSE
Peppermint	Increases attention & concentration
Citrus Oils (Orange, Lemon, Grapefruit, Lime, Tangerine, Yuzu, Bergamot)	• Improves mental clarity, focus • Orange acts as a mood booster • Lemon shortened reaction times • Bergamot reduces salivary cortisol
Sage	Prevents mental fatigue
Eucalyptus oils	Improves mental clarity, focus
Ylang Ylang	• Improves concentration • Induces calm

Basil	Mind-stimulation properties
Cedarwood	Induces dominance of Parasympathetic Nervous System to provide a calming, sedative effect
Rosemary	Enhances cognition
Rose	Induces relaxation
Lavender	• Improves the quality of sleep • Eases symptoms of anxiety
Marjoram	Reduces feelings of stress

(Sattayakhom, 2023).

The Antimicrobial Nature of Essential Oils

Providing a historical perspective of the antimicrobial nature of essential oils, the oils of oregano, rosemary, and marjoram were applied as skin antiseptics to support the healing of wounds. Their ability to be applied to the skin, absorbed, and internalized make them effective medicinal agents. Published literature indicates that essential oils possess a broad spectrum of antibacterial, antifungal, anti-viral, anti-parasitic, and anti-tumoral characters. We need to explore and exploit their use to address the modern-day enigma of bacterial resistance promoted by the excessive presence of antibiotics. Several studies support essential oil use against drug-resistant superbugs or bacteria which have adapted to resist antibiotics. Bacteria freely swap genes, reproduce quickly, and produce "infinite-like" populations, which promote resistance. Antibiotic-resistant bacteria exemplify survival of the most fit bacteria, as they survive the use of antibiotics to reproduce more colonies of resistant bacteria.

The emergence of antibiotic-resistance within bacterial populations has jeopardized our ability to fight infectious disease. However, the antibacterial action of essential oils offers an old, yet novel line of attack. The National Institutes of Health (NIH) conducted a study to determine the antibacterial and antifungal activity of some essential oils including; citronella, eucalyptus, geranium, lemongrass, orange, patchouli, peppermint, and palmarosa. The study was performed using 22 different bacterial species and 12 fungal strains. Lemongrass, eucalyptus, peppermint, and orange oils were effective against all 22 bacterial strains; while citronella, geranium, lemongrass, orange, palmarosa, and patchouli inhibited all 12 fungal strains (Pattnaik, 1996).

The unique combination of terpene, phenolic, ketone, alcohol, and aldehyde compounds in essential oils presents formidable inhibition against a wide-spectrum of microbes and it provides us pharmacologic potential. Whereas, bacteria have adapted to antibiotics, they have not adapted to the structural design of essential oils. Oils employ a different set of disruptive tactics when fighting bacteria. Oils destabilize bacterial cell structure with disintegration of the cell wall and the cell membrane. With fungal species, essential oils disrupt ATP or cellular energy production. Studies support that many viruses, such as herpes simplex, polio, and influenza *virus adeno* are negatively affected with application of essential oils (Tariq, 2019).

Simplistically, cancer results as a disruption of the cell cycle. As part of our genome, we have protective genes (tumor suppressors) which inhibit cancer formation; while we also have genes, which if mutated, over-stimulate the cell cycle. DNA mutations, specific to cell cycle regulation, lead to out-of-control cell division. Thus, cancer results in an over-proliferation of damaged, rogue cells having mutated DNA. In animal studies, essential oils are known to arrest cancer cells by different mechanisms. Constituent molecules of essential oils, such as linalool, limonene, geraniol, eugenol, beta-caryophyllene, carvacrol, thymol, and eugenol demonstrate cytotoxic character. This induces programmed cell death (apoptosis), or causes organelle dysfunction leading to cell death, as demonstrated in multiple cancer cell lines. Whereas, essential oils' cancer-inhibitive results have been demonstrated in cell lines or in animal models, very few clinical applications have occurred with human test subjects. Consequently, increasing the number of human studies with essential oils as anti-cancer agents is relevant (Sharma, 2022).

The Anti-Inflammatory Nature of Essential Oils

An apropos proverb reminds us, *"fire is never a gentle master,"* and inflammation acts as a fire within us. Many essential oils act in an anti-inflammatory capacity. Essential oils can suppress the immune response by regulating the release of inflammatory cytokines. This nature offers therapeutic potential to fight anti-inflammatory diseases, such as arthritis, auto-immune disorders, and cancer. The inflammatory response in the body is initiated in response to damage, such as a cut of the skin, the twist of an ankle, or the collection of plaque in blood vessel epithelial cells. Each would induce cytokine release resulting in acute inflammation. However, if the inflammatory response becomes chronic or over expressed then this presents as disease.

Traditional medicinal treatments to address pain associated with inflammation have included both steroidal and nonsteroidal agents. However, over-use in the short term and prolonged use in the long term may result in adverse health effects. NSAIDS (nonsteroidal anti-inflammatory drugs) were created to reduce inflammation and pain by inhibiting the action of enzymes known as cyclooxygenases (COX). The COX pathway initiates the production of prostaglandins, which play a role in the body's inflammatory response. NSAIDs include traditional pain and fever relievers such as, aspirin, ibuprofen (Advil), and naproxen (Aleve). The associated side effects of their use can be acute, as in the case of an upset tummy; however, overuse and prolonged use present increased risk of stomach ulcers and bleeding, heart attack and stroke, and reduced kidney function (Ghlichloo, 2023).

Topical application of anti-inflammatory essential oils can be used to address pain associated with acute inflammation. Inhaled frankincense or copaiba oils have demonstrated suppression of immune cell activity in the lungs, alleviating the symptoms of asthma. A study performed by Japanese researchers was conducted to analyze the effect of six different oils in suppression of the COX-2 pathway. The essential oils of thyme, clove, rose, eucalyptus, fennel, and bergamot were included in the study. Thyme oil with its active component, carvacrol, was found to suppress COX-2 expression by 75 percent (American Society for Biochemistry and Molecular Biology, 2010).

What other oils may be used to manage pain; thus, limiting our use of conventional medications? In a study of pain management for children after a tonsillectomy, children who inhaled the scent of lavender were able to reduce their daily dose of acetaminophen (Tylenol) after the surgical procedure. Another study showed that the scent of lavender was effective in lessening the severity of migraine headache symptoms. Additionally, when a peppermint gel was applied to the forehead, migraine pain was improved by at least one severity level. Bergamot was successful in reducing neuropathic pain, resulting from chronic nerve disease. Chamomile significantly reduced the need for pain relief in individuals suffering from osteoarthritis. Additionally, the oils of ginger, clary sage, eucalyptus, wintergreen, and rosemary have all shown effectiveness in addressing pain management for common ailments (Cronkleton, 2022).

The Immune-Supportive Nature of Essential Oils

Many oils are well studied in terms of their medicinal potential. Rosemary oil is one of those oils. Its historical use, especially to support cognition, goes back centuries. Its constituent profile is also known to benefit cardiovascular health. Rosemary's anti-inflammatory nature may promote brain health by preventing neuronal cell damage. Ongoing research is investigating its potential as an anti-inflammatory which is supportive for both cardiovascular and brain health, as well as its potential to inhibit cancer. Cinnamon bark oil has shown promising gastrointestinal effects. Its use as an agent to treat ulcerative colitis, irritable bowel syndrome, and dysbiosis will require rigorous randomized controlled trials. Metabolic syndrome is a combination of metabolic disorders characterized by obesity, insulin resistance, and hypertension. All of which contribute to the development of cardiovascular disease. Essential oil studies including research on the use of lemon balm, peppermint, lavender, bergamot, cypress, geranium, ginger, and black pepper as nutraceuticals, suppressed lipid droplet accumulation and adipogenesis; thus, exhibiting anti-obesity effects in the prevention of metabolic syndrome-related disorders (Chávez-Delgado, 2023).

Black seed oil is derived from the tiny black seeds of *Nigella sativa*. Its seeds have been used in traditional medicine for thousands of years. Black seed oil has demonstrated anti-inflammatory and antibacterial properties. Multiple studies indicate its immune-supportive character may help with conditions such as, psoriasis, asthma, wound healing, hair loss, and acne. Additionally, a 2021 study involving 55 participants with hypertension, showed most of the participants after consuming black seed oil experienced reductions in diastolic blood pressure, total cholesterol, LDL or bad cholesterol levels, and fasting blood sugar levels (Ajmera, 2024).

The pharmacologic potential of oils depends not only on their unique chemical compositions, but also on factors such as genetics, conditions under which plants are grown, proper harvesting times, and the use of standardized protocols in their extraction, including distillation and cold-press procedures (Pezantes-Orellana, 2024). Oil-producing plants grown under optimal conditions produce oils providing optimal results.

The following table provides a list of several oils with their protective, immune-supporting character which offer medicinal promise. The table elucidates the pharmacologic potential of essential oils; however, the pharmacologic potential is indeterminate, requiring a greater number of double-blind, placebo-based human trials.

Table 6: The "Anti" Nature of Essential Oils

Essential Oil	Protective Character	Application/Use
Lemon, Orange (Limonene)	Antimicrobial, anti-biofilm, anti-tumoral, antioxidant	Immune system supportive Food Preservation
Rosemary (Rosmarinic acid)	Antimicrobial, anti-viral, anti-tumoral, antioxidant	Inhibition of cancer proliferation Inhibition of neuronal damage
Thyme (Thymol, carvacrol)	Antioxidant, antimicrobial, anti-inflammatory, anti-tumoral	Protective against oxidative cell damage and cancer cell proliferation
Cinnamon (Cinnamaldehyde)	Antimicrobial, anti-inflammatory, anti-tumoral, anti-diabetic	Improved blood glucose levels Possible gastrointestinal support
Cilantro (Linalool)	Antiparasitic	Gastrointestinal support in treatment to inhibit nematodes
Black Seed Oil	Anti-inflammatory, antimicrobial	Anti-asthmatic, wound healing Reduction of blood pressure
Coriander (ß-linalool)	Antioxidant, antimicrobial, antibiofilm, antiparasitic	Gastrointestinal support in treatment to inhibit nematodes

Plant Immunity/Plant Defense

Because of their nature as molecules of scent, essential oils act in a defense capacity. Oils are emitted to reduce competition and limit predation. Herbivorous animals and insect pests thrive on plants, primarily above ground, while infectious microorganisms attack both above and below the ground. The inhibitory nature of essential oils on pest species is very useful. We do not have to spray noxious and toxic pesticides in and around the home, the yard, and the garden. For example, a spray made with peppermint oil is an effective ant deterrent. Ant hills often appear between stone walkways and paving stones. Any of the citrus oils can be sprayed onto the hills to disperse the ants from their habitat. Lemongrass and citronella oils are effective deterrents against mosquitoes. Once added to a carrier oil, they may be sprayed or topically applied to the skin. A common garden pest is the cabbage moth. A spray combining lavender oil with neem oil effectively deters the moth from alighting on a cruciferous plant. Rosemary oil mixed with neem oil effectively discourages cabbage moth larvae from consuming the plant. The oils can be added to a spray bottle containing either neem oil or witch hazel, water, and a liquid soap for better adherence on plant surfaces. The following identifies essential oils demonstrating insecticidal properties.

Table 7: Effective Essential Oils to Naturally Control Insect Pests

PEST IN THE HOME/YARD/GARDEN	SUGGESTED ESSENTIAL OIL(s)
1. **Ants**	Peppermint, Spearmint, Black Pepper
2. **Aphids**	Peppermint, Spearmint, Cedarwood
3. **Beetles**	Peppermint, Thyme, Oregano
4. **Cabbage Moth Caterpillars**	Rosemary
5. **Cabbage Moth**	Lavender
6. **Chiggers**	Lavender, Lemongrass, Sage, Thyme
7. **Flies**	Peppermint, Tea Tree, Sage, Rosemary
8. **Gnats**	Spearmint, Patchouli, Eucalyptus
9. **Mosquitoes**	Citronella, Lemongrass, Thyme, Pine

10. Slugs/Snails	Cedarwood, Pine, Hyssop, Patchouli
11. Spiders	Peppermint, Citrus oils, Lemongrass
12. Ticks	Geranium, Sage, Thyme, Tea Tree

(Petersen, 2024).

The Pharmacologic Potential of Essential Oils

The anti-factors of essential oils offer an array of pharmacologic and medicinal applications. Of course, these must be studied and scientifically validated to ensure safety of use and efficacy of the oils. Preliminary hospital settings have offered initial insight to this regard. Hospitalization and sleep often are not compatible. Seventy-two hospitalized patients were divided into two groups. One group received a gauze pad which had been laced with three drops of 40 percent geranium oil. The gauze pad was placed inside their pillow from 10:00pm until 6:00am. The control group received no intervention. Sleep quality was assessed via a validated questionnaire. The study indicated those who inhaled geranium oil had improved overall sleep quality as compared to placebo (Karimi, 2023).

Children also present poor sleep quality when hospitalized. A clinical trial evaluated the effects of inhalation of sweet orange essential oil. Children inhaled 2 drops of orange oil added to a gauze pad three times each day, 10:00am, 4:00pm, and before bed. The trial was conducted for three consecutive days. Inhaling orange essential oil resulted in improved sleep quality. It lessened the time for children to fall asleep, and they demonstrated less sleep disturbances through the night (Salafard, 2023).

While inhalation of oils is one delivery method for essential oils, a preclinical model was conducted on ingestion of copaiba oil. In this model, oral delivery of copaiba was studied for its use in the treatment of allergic asthma. Oral delivery did reduce inflammation and improved airway linings. Copaiba also may be inhaled through diffusion of the oil so it may be delivered to the upper and lower respiratory tracts. The study points to copaiba's promising use in restoration of respiratory airways for those who suffer from allergic asthma (Caputo, 2024).

The essential oil of rosemary has been investigated for its promotion of hair health. A pre-clinical model involving the topical application of 10 percent rosemary essential oil increased hair length, diameter of the hair follicle, and vascular endothelial growth factor or VEGF level. VEGF is integral to the flow of blood to hair follicles. In other words, the use of rosemary essential oil improved hair health (Khairnar, 2023).

Premenstrual syndrome is a monthly disturbance for girls and pre-menopausal women. It often involves bloating, breast tenderness and swelling, acne-flare-ups, mood swings, and abdominal pain. In a study performed on university students, women receiving lavender aromatherapy were found to experience less symptoms associated with premenstrual syndrome than women in the control group (Uzuncakmak, 2018).

In a preclinical model investigating wound healing, a 10 percent topical application of frankincense oil was investigated. Frankincense oil was found to reduce inflammation surrounding wound sites, which slows the healing process. Its application reduced cytokine release from injured cells and it reduced death of cells at the site of the wound. Moreover, frankincense promoted skin cell formation and enhanced collagen production at the wound (Venkatesan, 2025).

Rosemary oil was combined with Vaseline as an intervention for adult patients suffering from osteoarthritis. The ointment was administered twice daily for 10 days to patients receiving rosemary intervention. Placebo group received ointment of Vaseline, only. The control group received no intervention/ointment. Using a WOMAC score to measure pain, the group receiving the rosemary intervention reported greater alleviation of pain as compared to placebo and control group assessments (Zahra, 2022). The following table provides a summary of these studies and findings.

Table 8: Results of Pre-clinical/Clinical Trials Investigating Essential Oils

OILS STUDIED	HUMAN MALADIES	FINDINGS
Geranium oil, inhaled	Poor sleep, hospitalized adult patients	Sleep improvement
Orange oil, inhaled	Poor sleep, hospitalized children	Sleep improvement
Copaiba oil, ingested or inhaled	Allergic asthma	Reduced respiratory airway inflammation associated with allergic asthma
Rosemary oil, topical application	Hair loss, adults	Increased growth of hair
Lavender oil, topical application	Premenstrual syndrome, young women	Reduced premenstrual symptoms
Frankincense oil, topical application	Wound healing, adults	Reduced time for wound to heal
Rosemary oil, topical application	Osteoarthritis, elderly adults	Reduction of pain associated with osteoarthritis

A Biologist's Insight into Choosing Quality Essential Oils

Sustainability implies doing no harm to ourselves, our home, the wider environment and to limit our use of resources in doing so. **Less is more** is a different value system. Our new-found purchasing habits are made easier if we can find a company which embraces an analogous value system. We want a company that aligns with an environmental stance promoting sustainable practices and stewardship of the Earth. Good farmers stay in business, while bad farmers do not. We want a company that understands the quality of their oils corresponds with the quality of the soils upon which plants are grown. They should understand that the quality of oils from tree species is dependent upon the care taken in harvesting oils. Many essential oil companies cut corners by not valuing the oil production process from seed stocks, to planting, to growth and harvesting of plants, to distillation, to testing and even bottling. Sub-standard essential oil production produces sub-standard oils which provide sub-optimal results. In contrast, we want an essential oil company that values the essential oil process from seed to field to harvest to distillation to testing which delivers an optimal product providing optimal results.

SEED TO SEAL: The Young Living Essential Oils Promise (Essential Oils produced under optimal conditions provide oils giving optimal results)

A recent study reported in the British journal, *Nature Sustainability*, predicts a 50% decline in frankincense harvests in the next two decades. Threats supporting such a precipitous decline include: over-grazing by domesticated livestock, insect predation, human-induced brush fires, and domestic political unrest. *Boswellia* (frankincense) species in Ethiopia, Sudan, Oman, Yemen, India, Somalia, and Kenya are experiencing these threats, resulting in increased mortality in established trees. Consequently, adult trees are being over-tapped, while young trees are not making it to tappable-age. Authors of the study warn, "*Concerted conservation and restoration efforts are urgently needed to secure the long-term availability of this iconic product.*" (Bongers, 2019).

Young Living recognizes such threats in maintaining its farms around the globe. In the country of Oman, Young Living works with the government to develop frankincense reforestation tracts. Ancient harvesting methods are relied upon to produce sustainable frankincense. For example, each frankincense tree can be harvested only once in a 12-month time frame. Young Living's Palo Santo oil is harvested from only fallen or dead trees located on its *Finca Botanica Aromatica* farm in Ecuador. Fallen Palo Santo trees remain on the ground for two to four years before harvesting occurs. Once trees are harvested, the oil is extracted from the trunk and roots. Similar conservation practices apply with Young Living's partnership with the Kona Sandalwood Reforestation Project in Hawaii. Only the dead wood from sandalwood trees is harvested to produce Royal Hawaiian Sandalwood oil. Additionally, sandalwood seedlings are planted with companion trees as part of a 40,000-acre reforestation project (youngliving.com/us/farms/2025).

Young Living places its soils before its oils. The quality of the plant and the oil it produces begins with the nurturing substrate upon which it is grown. Soils deficient in quality of nutrients, structure, and water-holding capacity produce sub-standard plants and oils. Young Living stringently follows their proprietary **SEED to SEAL** protocol, which utilizes a 3-pronged emphasis: **sourcing, science, and standards**. This ensures production of oils of the highest quality, while employing a sustainable emphasis throughout the oil supply chain. With Young Living, a correlation exists between soil methodologies implemented at their farms aligned with best soil conservation practices. These practices include the following:

1) **Farms located in rural locations away from urban pollution sources**
2) **Exclude the use of synthetic soil amendments (pesticides/herbicides/fertilizers)**
3) **Employ no-till or limited till cultivation**
4) **Practice crop rotation**
5) **Plant Nitrogen-fixing cover crops as a pre-prep soil amendment**
6) **Use Nitrogen-fixing cover crops rotationally for continual soil improvement**
7) **Addition of organic compost and other green manure amendments**
8) **Irrigate using natural springs, aquifers, deep water wells, and mountain snow melt reservoir**

Young Living's agronomists and farm managers recognize what is taken from the soil must be placed back into the soil; this is sustainability (Essential Oil Desk Reference).

Young Living embraces the use of many sustainable practices on their farms. To illustrate, when juniper is harvested at their Whispering Springs Farm (Mona, Utah), it is shredded prior to its placement in distillery vats. After distillation of juniper oil, plant residue remains which is then used as compostable green manure. Young Living's *Finca Botanica Aromatica* Farm (Guayaquil, Ecuador), maintains a multi-rowed worm house composed of worm bins. Worm bins produce two organic products, a liquid tea fertilizer, and worm castings, which are both used to enrich soils. Young Living's many lavender farms use organic fertilizers composed of green manure, worm castings, and worm bin tea. Weeds are controlled with the use of mulch, hoeing, hand pulling, and sheep grazing. An organic weed and pest spray combines essential oils with castile soap.

Attention to farm location, soils, and water resources is another critical step to ensure every Young Living essential oil and oil-infused product is free of synthetic toxins, heavy metals, and other environmental contaminants. Seed lines are selective, properly classified, and from original seed stock. They are planted with care, germinated under controlled conditions, and hardened to outdoor conditions before planting in fields. Field planting is species specific to ensure optimal timing for growth and harvest. Optimal harvest time accounts for the condition of the plant, time of year, even time of day. For example, melissa or lemon balm oil is best harvested just before or during initial flowering when the highest concentration of volatile oils is available.

Having distilleries built on-site is crucial for optimal distillation of oils. Some plant species must be distilled immediately after harvest, while other plants may age in the field to achieve peak oil value. Young Living combines stainless steel technologies with old world distillation practices in the distillation process. Stainless steel vats reduce the potential for corrosive contaminants. Instead of employing fast track extraction protocols, which encourages sub-par oils, Young Living extracts oils using low pressure and low temperatures over longer extraction times. Other oil companies use quick extraction, distilling plants at higher temperatures and pressure. Denaturation of the oil can occur under these conditions. Economically speaking, quick production is cheaper resulting in the oil's price being cheaper, but at the cost of an oil cheapened in quality. Young Living's Seed to Seal protocol values the integrity of the oil and its unique chemistry. No oil is produced before its time from plants in the field, to harvest, to extraction/distillation.

Each of Young Living's oils must pass stringent testing to ensure purity. Just as human fingerprints are unique, an oil's chemistry is unique, as identified by its molecular minutiae. Young Living uses its own internal labs, as well as third-party audits, to verify that international oil purity standards are met. Oil purity is assessed and verified by using analytical techniques such as, gas chromatography/mass spectroscopy (GC-MS), high performance liquid chromatography (HPLC), and isotopic ratio mass spectrometry (IRMS). These analytical techniques authenticate an oil's purity and if any adulteration of the oil has occurred. Testing, testing, and more testing determines whether a Young Living oil will ever make it into a customer's home. For the consumer, it is the ultimate insurance policy, the trust factor is attained (youngliving.com).

Young Living's proprietary Seed to Seal protocol authenticates the oil's quality. As consumers, we want to purchase pure, unadulterated oils to receive the full efficacy of their nature. To illustrate, adulterated oils may result from contamination during the distillation process or by growing plants under non-optimal conditions. This counters the rationale for the purchase and use of essential oils in the home. Young Living's premium quality, unadulterated essential oils are found in a diverse range of consumer-based products. The products are plant-derived, naturally biodegradable, safe for the consumer, and the wider environment (www.youngliving.com). Ethical standards and trust in place, Young Living's essential oils and oil-based products provide the consumer safe alternatives for replacement of household toxic products, **one swap at a time.**

From a biological and a scientific perspective, my trust in Young Living Essential Oils resides in their attention and regard for following best practices. This includes: maintaining their own seed lines, employing best practices in the growth, harvesting, and distillation of their oils, utilizing rigorous state of the art testing to ensure their oils are of the highest quality, and their commitment to following sustainable practices throughout the oil production process. Oils produced under optimal conditions produce oils providing optimal results; this is the Young Living standard.

Sustainability: Putting "Less Is More" Into Practice

A conservative lifestyle, such as "less is more," aligns with sustainability by encouraging us to be mindful consumers/shoppers. Initially, this manuscript alerted you to synthetic chemical toxins and their all-to-common presence in widely available household, consumer-based products. As aware consumers, we must initially identify their toxic presence in our home. Next, we must become more mindful with our purchasing habits to not repeat their return. Mindful shopping, mindful steps for removal, and mindful replacement with safer, healthier alternatives makes for a less toxic, restorative home. For every toxic product that we find an equivalent non-toxic replacement, with its purchase and its use, we achieve one less toxic exposure. Being a smarter consumer promises a safer, more restorative environment for our family, our community, and the Earth. Mindful consumers practice the following to make them habit (adapted from Toxic Matters, 2013).

1. **Reconsider a definition of beauty, in terms of the use and purchase of personal care and cosmetics.**
2. **Avoid purchase of fragranced or scented personal care products.**
3. **Take caution with cleaning product purchases (counter, floor, window, dishwashing, bathroom, clothes detergent, dryer-products, etc.)**
4. **Diffuse essential oils and do not purchase air fresheners and petroleum-based candles.**
5. **Take caution with cookware and kitchen utensil purchases. Avoid Teflon purchases.**
6. **Do not spray pesticides/herbicides in and around the home.**
7. **When doing home renovations, ask for VOC-free, water-based products and with furniture purchases allow products to off-gas, off-site.**
8. **Purchase clothing made of natural fibers, cotton, wool, linen, silk, bamboo, and beware of water-resistant/stain-resistant fabrics.**
9. **Use essential oils as first choice when common ailments arise.**
10. **Dust and wet mop using essential oils on a regular basis to address toxin-laden dusts in the home.**
11. **Open household windows (seasonal basis).**
12. **Remove shoes before entering the home (do not walk toxins in).**
13. **Encourage frequent hand washing with oil-infused soaps, especially before eating. Make it a habit.**
14. **Avoid the purchase of plastics, #3, #6, and #7, the non-recyclable plastics (#3 is PVC, #6 is polystyrene/Styrofoam, and #7, BPA-problematic).**

15. **Never microwave or heat foods in plastic containers (toxins leach).**
16. **Plastics are not dishwasher-safe (toxins leach).**
17. **Avoid foods packaged in plastics at the store (toxins leach).**
18. **Reduce the contact of foods in plastic containers/wraps at home (toxins leach).**
19. **Reduce the purchase of ultra-processed foods (products made with industrially produced ingredients and additives, such as colorants and preservatives).**
20. **Limit the purchase of canned and processed foods (BPA contamination).**
21. **Consume plant-based, whole, nutrient-dense foods.**
22. **Shop the organic section of the grocery and purchase locally-sourced foods.**
23. **If supplementation is necessitated, choose oil-infused supplements.**

NIH investigator, Dr. Germaine Buck-Louis, a reproductive and child developmental specialist, concludes that actions, such as the ones just listed make a difference. Research studies indicate people with a change in purchasing habits aligned with selective lifestyle change, show greatly reduced levels of toxins in a manner of days. She emphasizes, "*Lifestyle has an important place here.*" (Buck-Louis, 2016.). Adapt, adopt, embrace change with the purchase of alternative, healthier, consumer-based products providing a less toxic, healthier household environment for you and your family.

Less Is More, One Swap at a Time

As a biologist, I suggest the adoption of a Young Living Essential Oils lifestyle to achieve a healthier, safer, more restorative home environment. Young Living offers a wide array of essential oils, oil-based products, oil-enhanced nutritional products, and dietary supplements. Young Living's proprietary and rigorous **SEED to SEAL** process ensures oils that are not adulterated. Consequently, Young Living's Essential Oils are of the highest quality, purity, potency, and efficacy. Oils produced under optimal conditions produce oils providing optimal effects/results. Young Living combines their oils with plant-based, natural ingredients in the production of consumer-based products that offer us alternatives to commercial-based more toxic-laden products. With Young Living, a less toxic, restorative home is truly attainable.

Young Living's THIEVES Line of Products (Reduce Toxic Exposures)

The household cleaning process needs to reduce toxic exposures, while providing a refreshing, hygienic cleaning experience. There is no "greenwashing" with Young Living's Thieves line of products. Rest assured, these products are formulated with biodegradable, plant-based and Earth-friendly ingredients. The ingredients formulated in Thieves concentrate include: water, alkyl polyglucoside (natural sugar-based surfactant), sodium methyl 2-sulfolaurate (coconut oil surfactant), disodium 2-sulfolaurate (coconut oil surfactant), tetrasodium glutamate diacetate (plant-based chelating agent), and Young Living's Thieves oil blend.

Thieves oil, the basis for the concentrate and all other products in the Thieves line, is a blend of clove, cinnamon bark, lemon, rosemary, and *Eucalyptus radiata* oils. Because plants utilize essential oils for microbial defense, the oils naturally are antibacterial, antifungal, and antiviral in character. This is supported by multiple research studies. According to Dr. Robert Fuller, *"Plant biochemists term these oils or plant extracts as "allelopathic" agents. These are part of the plant's natural defense mechanisms allowing them to ward off many of the same pathogens which are of concern to us. It is no wonder they have such profound medicinal properties as well, and they have been tested by mankind for thousands of years. These include the aromatic plants that are familiar to us because of their pleasant scents, such as members of the cinnamon, rosemary, or clove families. This natural defense of each plant has health implications for us. Essential oils with their protective properties are supportive of our body systems."* (Fuller, 2025).

The Thieves blend of clove, cinnamon bark, lemon, rosemary, and *Eucalyptus radiata* oils synergistically work together to make Thieves a powerful and diverse antimicrobial agent. To illustrate, vinegar (pH value of 2) is limited in the diversity of microbial species than can be inhibited by its single-note acidic nature. Alternatively, Young Living's Thieves blend of oils potentially inhibits a broader-range of biological microbes, providing greater anti-pathogenic potential. Because Thieves Cleaner is prepared as a concentrate, it can accomplish a variety of cleaning tasks from general cleaning to the removal of oil stains from the garage floor. Its ultra-concentrated formula is economical, since it can be diluted to address any cleaning task. Follow bottle directions. A safer, healthier, less toxic environment is created with the use of Thieves and a list of its benefits follows:

1. **Plant-based formulation makes it biodegradable which reduces toxic potential of the home**
2. **Non-abrasive, cleans and rinses without leaving a residue**
3. **Fresh, inviting sweet/spicy, invigorating scent**
4. **Directions for use to address a diversity of cleaning needs**
5. **Economical, one capful added to 2 cups of water is required for general cleaning. A little bit of Thieves Cleaner goes a long way to meet your general cleaning needs.**
6. **Suitable for counters, walls, flooring, carpet, drapes, upholstery, clothing and bedding fabrics, glass, etc.**
7. **Safe for septic systems and waterways**
8. **The THIEVES Product Line includes a variety of cleaning, personal care, and oral care items available to replace toxic products in the home.**

Table 9: Young Living's THIEVES Product Line

Cleaning Products	Personal Care Products	Oral Care Products
• Household Cleaner	• Waterless Hand Sanitizer	• AromaBright Toothpaste
• Dish Soap	• Cleansing Soap	• Whitening Toothpaste
• Fruit & Veggie Spray • Fruit & Veggie Soak	• Easy Carry Spray	• Fresh Essence Plus Mouthwash
• Laundry Soap	• Chest Rub	• Hard Lozenges
• Foaming Hand Soap	• Thieves Oil Blend	• Thieves Mints
• Dishwasher Tablets	• Disinfecting Wipes	• Infused Cough Drops
• Multi-purpose Wipes	• Thieves Oil Blend and Thieves Roll-on	• Dental Floss
• Kitchen & Bath Scrub		

*youngliving.com provides a complete catalog of their essential oils and oil-enhanced products

Table 10: Young Living's THIEVES Product Line Reduces Toxic-exposures

Less Is More One Swap at A Time

Consumer-based products	Toxic-exposures	Young Living's toxin-free replacement answer
Cleaners	Phthalates (EDC) Ammonia Chlorine bleach (EDC) Formaldehyde preservative, VOCs Sulfates (SLS, SLES) Synthetic fragrance Caustic Lye	**Thieves concentrate** limits exposure to toxins for you and the environment.
Dish Soap	Phthalates (EDC) Sulfates (SLS, SLES) Formaldehyde preservative Synthetic fragrance	**Thieves Dishwashing Products** limits exposure to toxins for you and the environment.
Laundry Soap	Phthalates (EDC) Chlorine bleach (EDC) Sulfates (SLS, SLES) Formaldehyde preservative Synthetic fragrance	**Thieves Laundry Soap** limits exposure to toxins for you and the environment.
Foaming Hand Soap	Sulfates (SLS, SLES) Parabens (EDC) Formaldehyde preservative	**Thieves Foaming Hand Soap** limits exposure to toxins for you and the environment.
Hand Sanitizer	Methanol, propanol	**Thieves hand sanitizer** limits exposure to toxins.
Bar Soap	Phthalates (EDC) Parabens (EDC) Sulfates (SLS, SLES) Synthetic fragrance	**Young Living's Bar Soap** limits exposure to toxins for you and the environment.
Toothpaste	Sulfates (SLS, SLES) Parabens (EDC) DEA (diethanolamine) Heavy metals, artificial colors	**Young's Living's AromaBright and Whitening Toothpastes** limit exposure to toxins for you and the environment.

Young Living's Oil-Enhanced Personal Care Line of Products

Young Living's extensive personal care line of products is infused with premium quality essential oils. In alignment with their emphasis on sustainability, Young Living's personal care products are essential-oil enhanced, plant-derived, biodegradable, and **especially formulated without the presence of EDCs**. Additionally, they avoid the use of sulfates, triethanolamine, talc, synthetic surfactants, parabens, artificial coal tar dyes as colorants, and other suspect chemicals. Natural surfactants include: glucosides, soapberries, and quillaia extract. Preservative compounds include: potassium sorbate and sodium levulinate salts, and citric acid, a preservative which slows oxidation. Their goal in the formulation of each oil-enhanced personal care product is to provide products which reduce exposure to harmful chemicals.

Table 11: Young Living's Oil-enhanced Personal Care Products

Less Is More One Swap at A Time

Consumer-based products	Toxic exposures	Young Living's Toxin-free Replacement Answer
Hair Care: shampoos, conditioners	Sulfates (SLS, SLES) Parabens (EDC) Phthalates (EDC) Formaldehyde preservative Synthetic fragrance	Lavender Volume Shampoo and Conditioner Copaiba Vanilla Shampoo and Conditioner
Facial Care: cleansing oil, facial scrub, creams, acne treatment, lip balms	Phthalates (EDC) Parabens (EDC) Sulfates (SLS, SLES) Formaldehyde preservative Oxybenzone (EDC) Synthetic fragrance	Balance Product Line Bloom Products Satin Facial Scrub Sandalwood Boswellia Firming Wolfberry Eye Cream Acne Treatment Lip balms
Body Care: deodorants, sunscreen, skin sprays, pain cream, sports gel, insect repellent, shaving products	Oxybenzone (EDC) Formaldehyde preservative Parabens (EDC)	Mineral Sunscreen LavaDerm & ClaraDerm Sprays Cool Azul pain cream, sports gel

	Phthalates (EDC) Heavy metals Coal tar dyes Sulfates (SLS, SLES) Synthetic fragrance	Insect Repellent Aroma Guard Meadow Mist & Mountain Mist Deodorants Beauty Glow with Full-Spectrum CBD
Moisturizers and lotions	Parabens (EDC) Phthalates (EDC) Sulfates (SLS, SLES) Formaldehyde preservative Synthetic fragrance	Lavender, Genesis, Sensation, & Lucious Lemon lotions Regenolone Cream Prenolone Plus Body Cream Rose Ointment Body Butter Collection
Soaps and body gels	Parabens (EDC) Phthalates (EDC) Formaldehyde preservative Synthetic fragrance	Gels (Evening Peace, Morning Start, Sensation, Dragon Time, Lavender) Essential Oil Soaps

*youngliving.com provides a complete catalog of their essential oils and oil-enhanced products.

Young Living's Just for Kids Line of Products

Our children are our legacy (what we leave behind); while at the same time, they are the future (what we leave ahead). Our modern-day, consumer-based society, effusive with the presence of synthetic chemical toxins, is not a nurturing environment. This makes our children's exposure to household toxins quite precarious. Let us make sure that a reduction in their exposure to synthetic chemical toxins in the home occurs expeditiously. Young Living is cognizant of providing our children a healthier, less toxic path with the KidScents and SEEDLINGS product lines. Consider the following.

Table 12: Toxic Exposures, Our Children, and Young Living's Replacements

Less Is More One Swap at a Time

Household Location with possible toxic additions	Toxins to Avoid	Young Living's Non-toxic Products
Nursery: • Wipes • Diaper Cream • Baby Lotion • Baby Oil	Parabens (EDC) Phthalates (EDC) Formaldehyde preservative Synthetic fragrance Flame Retardants (textiles)	**Seedlings** Baby Wipes, Diaper Cream, Baby Lotion, and Baby Oil are formulated with essential oils. They are tear free, toxin-free, and gentle for use with baby.
Bathroom: • Shampoo • Conditioner • Body Wash • Baby Lotion	Parabens (EDC) Phthalates (EDC) Sulfates (SLS, SLES) Formaldehyde preservative 1,4-dioxane Synthetic Fragrance Bisphenol A (BPA)	**KidScents** Line of Products and **Seedlings** Products are naturally-derived, botanically - enriched, toxin-free, and formulated with essential oils to be gentle, apply with ease, and skin hydrating with every use.
Laundry Room: • Laundry Detergent • Dryer Sheets	Fragrances & Dyes Optical Brighteners	The **THIEVES** Line of Products are formulated from naturally derived plant-based

• Fabric Softeners	Formaldehyde preservative Sulfates (SLS, SLES) Phthalates (EDC) Parabens (EDC)	formulations. They are essential oils-enhanced. Oil-enhanced wool drier balls, in lieu of use of dryer sheets.
Outdoors: • Sunscreen • Bug Spray	DEET Oxybenzone (EDC) Octinoxate (EDC) Synthetic Fragrance Parabens (EDC) Phthalates (EDC) Retinyl Palmitate (vitamin A)	Young Living's (SPF 50) **Mineral Sunscreen Lotion** is reef safe and free of harsh chemicals. Young Living's **Insect Repellent** is a natural, oil-infused, lotion to deter insect pests. Young Living's Insect Wipes
Kitchen: • Cleaning products • Kitchen utensils • Plastics and Food • Nutrition	Bleach (EDC) Ammonia Phthalates (EDC) Parabens (EDC) VOCs	The **THIEVES** Line of Products KidScents Nutrition Support: • MightyVit • MightyPro

*youngliving.com/products/2025

It is of particular importance to highlight two consumer-based products that are **applied to the whole body** of a child. Additionally, consider what the conventional consumer product offers in terms of toxic exposure. In a recent study conducted at University of California (Davis), urine samples collected from toddlers showed increased levels of benzophenones (EDC) from sunscreens, and pyrethroid pesticide residues from insect repellents. Their mothers also had provided urine samples during pregnancy. Comparing mother to child urine samples, the children had higher levels of several chemicals, including two phthalates, bisphenol S (replacement for BPA), and constituents of pesticides (Oh, 2025).

As consumers, what goes on, eventually rinses off; case in point, the use of sunscreens. Once slathered with sunscreens, we slough-off approximately 14,000-25,000 tons of commercially-available sunscreens into fresh and saltwater ecosystems each year. Two ingredients, in particular, oxybenzone (BP-3), and octinoxate, are toxically problematic, especially rinsed off into ocean waters nurturing the development of coral reefs. Be reminded both substances are recognized endocrine disruptors (EDCs). These toxins exacerbate the phenomenon of coral bleaching, leading to the demise of the reef. Oxybenzone and octinoxate damage coral cell DNA, expressed by abnormal growth and development in the coral colonies. Even though, in some cases, their presence is banned regionally, oxybenzone is found in more than 3,500 sun-protective skin care products manufactured around the globe. As recognized EDC compounds, **why would we want to slather it all over the bodies of our babies and children?** (NOAA, 2024).

In contrast, Young Living produces a Mineral Sunscreen Lotion (SPF 50) that contains neither oxybenzone or octinoxate. Instead, its active ingredient is non-nano zinc oxide (22 percent), which reflects harmful radiation away from the user, so rays do not penetrate the skin. By including in the formulation, the essential oils of helichrysum, lavender, myrrh, cistus, carrot seed, and frankincense, greater sun protection is achieved. As a broad-spectrum lotion, it protects against UVA rays responsible for accelerated aging of skin, and mutagenic UVB rays. The lotion is dermatologist-reviewed, hypoallergenic, non-greasy, and both water and sweat resistant (www.youngliving.com).

Sunscreen (SPF 50) is not a trivial product in the prevention of the development of skin cancer. No matter the skin type, all of us should apply a skin-protective sunscreen varying in SPF (sun-protective factor) from 10 to 50. The Skin Cancer Foundation recommends a water-resistant, broad-spectrum sunscreen with an SPF of 30 or higher, especially for those having the most vulnerable skin type. Additionally, we can protect ourselves with common sense measures such as covering up with light-colored, breathable fabrics, wearing wide-brimmed hats, UV-blocking sunglasses, and staying out of the sun between the hours of ten o'clock to two o'clock. There is no such concept as a healthy tan.

With Young Living's SPF 50 Mineral Sunscreen Lotion, our children playing in the sun and sand, while also saving baby corals, does not have to be a mutually exclusive occurrence. Being a mindful consumer with an eye on sustainability is a win for you, a win for your family, and a win for all of Earth's inhabitants, like baby corals.

Let us consider one other personal care product that we often spray or apply all over our bodies and the bodies of our children. As exemplified by the previous sunscreen example, we value our time out of doors, especially in the warmth of summer's sun. Taking the good with the bad, we apply insect repellents to keep away mosquitoes, ticks, flies, and other biting insects. Insect repellents are regulated as pesticides in the United States. They contain active ingredients which must be listed on labels, while inert or inactive ingredients do not have to be listed. DEET or N, N-Diethyl-meta-toluamide, is the primary active repellent agent in most commercially available insect repellents. DEET exposure is linked to the following acute responses: skin irritations, headaches, and shortness of breath (problematic for asthmatics). It is easily absorbed across the skin. It enters soils and surrounding waterways, where it breaks down slowly. DEET has been detected in groundwater, surface water, and drinking water resources ("Product Profile: Bug Repellent." 2023).

Also problematic are pyrethroid and permethrin additions to insect repellents. There are 1,000-plus pyrethroid insecticides. Pyrethroids are lipophilic; therefore, they interact with fat-loving tissues, such as human skin. They easily enter the body where they have been found to cross the blood-brain barrier. Pyrethroids are associated with endocrine disruption. Acute responses include: irritation to the skin, initiation of the asthma response, nausea, and burning/itching of the skin. Permethrin is a synthetic pyrethroid compound which is often used in the manufacture of outdoor gear and clothing. It is an insecticidal agent which kills insects on contact. Thus, it is harmful to important insect species, such as honey bees and other insect pollinators.

Young Living's Insect Repellent is composed of 100 percent naturally-derived, plant-based ingredients. It is formulated without DEET, parabens, filler, phthalates, petrochemicals, synthetic preservatives, or synthetic fragrances and colorants. Its formulation in a sesame oil base, includes the following essential oils: citronella, lemongrass, rosemary, geranium, spearmint, thyme, and clove. This combination of oils is used to mask human skin odors; thus, affecting an insect's sense of smell and taste. Young Living's oil blend emits a pleasant, citrusy scent for the user. It easily rubs onto the skin without leaving any greasy or sticky residue. Its hypoallergenic formulation leaves the skin feeling smooth and moisturized. When applying, care should be taken around the eyes and mucous membranes. Apply to the hands first and then take care with hand application to facial areas. Additionally, we can use common sense measures to prevent insect pests from alighting and biting. Wear light colored, breathable fabrics, which cover arms and legs, so repellents are applied only to uncovered skin areas.

With Young Living's natural, plant-derived, biodegradable ingredients, there is no harm with application to you or your family members, and no environmental harm to insect pollinators and aquatic organisms inhabiting ground or surface water resources. Another example of a win for you, a win for your family, and a win for Earth's inhabitants. **Putting less is more into practice is sustainability.**

Oil-Enhanced Nutrition with Young Living's Nutritional Supplements

Hippocrates receives credit for the following, *"Let food be thy medicine and medicine be thy food."* The ancient doctor was linking wellness with supportive nutrition. Nutrition is integral for maintenance of homeostasis in the body, as we require calories, protein, fats, vitamins, and minerals to maintain body structure and metabolism. Prior to the 20th century, we were much closer to our food sources than we are today. Rural living meant growing one's own food, while city folk bought foodstuffs from a locally-sourced market. The closer we are to the source of our food, the fresher, and more nutritious it will be.

Modern-day, big business agricultural practices favor food yield at the expense of food nutrient levels. In support of this business model, soil nutrients have been depleted with the continual application of synthetic fertilizers, pesticides, and herbicides (Davis, 2004). Fruits and vegetables are harvested before they reach maturation and peak nutrition. They are boxed onto trucks to mature over the long haul, while losing nutrient value along the way. Fruits and vegetables are 'shellacked,' to enhance their appearance on grocery shelves. Beware of the waxy feel and extra shine on tomatoes, peppers, and apples.

Food processing on a grand scale exacerbates nutrient loss even more. Seventy percent of the American food supply is composed of ultra-processed foods, making their consumption difficult to avoid. Ultra-processed foods include a broad range of ready to eat products, such as packaged snacks, carbonated beverages, instant noodles, and ready-prepared foods. These industry-prepared foods are rife with additives to enhance color, flavor and taste, texture, and durability; thus, they are more fake than food. Avoidance is required, as these dietary offerings predispose consumers to a greater likelihood of cardiovascular disease, and age-related neuro-disorders, such as Parkinson's disease (Miller, 2025).

Ultra-processed foods dominate the American food supply with 50 percent of our energy intake coming from this source. Contemporaneous with the physical maladies resulting from their increased intake, is increased likelihood of adverse mental health outcomes, such as those demonstrated by patients suffering dementia. Ultra-processed foods are identified as a source of microplastics ingestion. Nano-shards of plastics have been noted in brain, lung, intestine, liver, and kidney tissues (Nihart, 2025). The toxic potential of our processed food supply results from the use of synthetic fertilizers, and pesticide/herbicide applications. Then there is the packaging of foods in plastic, the canning of food with lids possibly lined with BPA, the addition of synthetic colorants, and the presence of questionable preservatives. All contribute to a food supply having anemic nutrition, while also being toxically tainted.

Essential oils derived from herbaceous plant species and citrus fruits can help to address nutrient-inadequate foods. This is especially true if the oils are distilled or cold pressed using stringent protocols to insure purity, potency, and efficacy. Pure oils may be ingested as foods. In fact, some essential oils have received FDA-GRAS (generally recognized as safe) designation for use in food. The following food-grade oils have received the FDA's blessing: peppermint, lemon, orange, lavender, ginger, cinnamon, and basil. Thus, essential oils can be used, in lieu of herbs or spices, to enhance flavor in cooking, to calm the gastrointestinal system, and to enhance nutrient uptake and bioavailability (Sparks Akers, 2024).

Just as we consume herb species, we can ingest pure essential oils, in minute quantities. For example, one drop of pure peppermint oil equates to drinking approximately 25 cups of peppermint tea. Because of their chemical profiles, essential oils may be ingested in concert with a carrier-delivery oil, such as coconut, sweet almond, or olive oil. Oils can complement the digestive process. A meta-analysis review of peppermint oil as a treatment for abdominal pain associated with irritable bowel syndrome (IBS) found vegetable-coated peppermint oil capsules to be a safe and effective therapy for relief of associated symptoms (Alammar, 2019). Antiemetic ginger oil inhibits the brain's nausea center; thus, it soothes and calms the gastrointestinal tract. Several animal-based studies indicate that essential oils in feeds improve the microbial balance of the intestines with improved nutrient absorption. Additionally, essential oils were shown to stimulate the production of digestive enzymes; thus, improving digestion and nutrient absorption (Abd El-Hack. 2022).

Acknowledging that our food supply is nutritionally lacking warrants supplementation to aid in filling nutritional gaps. Certain criteria such as, age, health conditions, life stages (pregnancy, nursing, picky eaters), and food choices further impact nutrient uptake. While supplements are not a substitute for a healthy diet, they complement a diet by bolstering its nutrient profile.

On the grand scale of food supplementation, food manufacturers reduced the incidence of rickets with vitamin D fortification of dairy products. Folic acid fortification of cereals and grains diminished the incidence of neural tube defects; while making iodized salt available reduced the incidence of goiter. Consequently, for those who suffer cardiovascular conditions, fish oil, and vegan supplements, rich in omega-3 fatty acids, are beneficial. A supplement formulation high in vitamins C and E, copper, zinc, beta-carotene, lutein, and zeaxanthin, is recommended for individuals diagnosed with age-related macular degeneration (AREDs). Since the B vitamins are not produced by the body, we must ingest them through diet, mainly with the consumption of animal proteins. If you are not a meat lover, then B vitamin supplementation is warranted. An additional, unfortunate consequence of the aging process is reduced nutrient absorption, so supplementation is suggested as we age (ods.od.nih.gov/factsheets).

As a biologist, I have found one of Young Living's products particularly checks many nutritional boxes. Young Living's NingXia Red is a whole fruit wolfberry puree combined with juices of aronia, sweet cherry, plum, blueberry, and pomegranate. This puree is enhanced with essential oils of orange, tangerine, lemon, and Yuzu. Two 2-ounce servings per day gives a nutrient infusion which delivers whole body support, specifically targeting eye, digestive, and cardiovascular health. By utilizing the whole wolfberry superfruit, NingXia Red provides fiber, eye-healthy zeaxanthin, polysaccharides, 18 amino acids, 21 trace minerals, and vitamins B1, B2, B6, and E. NingXia Red is restorative because it replenishes key nutrients, negates the effects of free radical disruption or oxidative stress in the body, and has a low glycemic index.

Fat-soluble vitamins, A, D, E, and K are better delivered, absorbed, and utilized by the body if delivered in combination with an oil base. Dr. Ashley Cribb, promotes oil-infused supplementation because essential oils act as biochemical catalysts, assistive to the delivery process. Dr. Cribb writes, *"All vitamins and minerals are more easily assimilated into the body in the presence of enzymes and essential oils." (Cribb, 2022).* We can observe this assimilative potential in Young Living's NingXia Red formulation. The combination of the wolfberry/fruit juice puree combined with the citrus essential oils makes for better nutritive delivery and assimilation at the tissue and cellular level.

The human gut microbiome is being scrutinized for the prominent role it plays in our physical and mental wellbeing. Research studies strongly support the existence of a gut-brain connection. Gut bacteria produce neurotransmitters such as serotonin, dopamine, and GABA, which affect mood. The varied bacterial groups making up the gut's microbiome are integral in maintaining human metabolism; while also acting as a communication conduit between the gut and the brain. This would suggest that gut bacteria play a role in regulation of homeostasis in the body. Additionally, it is important to maintain the balance of organisms occupying the gut microbiome. Young Living provides prebiotic and probiotic supplements in support of the microbiome. Since 70 - 80 percent of our immune system resides in the digestive system, a healthy, supported gut is crucial for our overall health and wellbeing (Cohen, 2021).

Turmeric and its active ingredient, curcumin, has long been recognized as beneficial for health. It is the spice added to many Indian foods providing the yellow color notable in curry. Multiple research studies indicate turmeric presents anti-inflammatory, antidiabetic, and antioxidant properties (Gunnars, 2023). Oxidative damage in the body involves the release of highly reactive free radicals. Because these molecules have unpaired electrons, they react with organically stable molecules, such as fatty acids, proteins, and DNA, to disrupt their stability. Consequently, curcumin may play a role in negating the effects of the production of free radicals and the role they play in the aging process and the manifestation of many diseases.

Because curcumin is fat soluble, it is best to combine it with a fat emulsion, such as essential oils. Young Living's Golden Turmeric is formulated with soluble tapioca fiber, *Boswellia serrata* (frankincense) resin extract, ginger, and lime essential oils. This combination makes the curcumin 24 times more bioavailable than standard turmeric extract. Young Living's Golden Turmeric with its mango rose flavor combines the benefits of curcumin with essential oils to naturally support the immune response, joint health, and our ongoing battle with inflammation.

Young Living's NingXia Red

This Photo by Unknown Author is licensed under CC BY-SA-NC

Turmeric (curcumin) pictured below

This Photo by Unknown Author is licensed under CC BY

Following is a table of Young Living's oil-infused supplements which can be taken to address nutritional gaps. Unless one consumes an organically-based whole-foods, non-processed-foods diet, nutritional gaps result. The aging process exacerbates nutritional gaps. Supplementation is suggested to maintain the correct balance of microbial species (good bacteria versus bad bacteria) in the immune-supportive gut. Supplements infused with essential oils act as catalysts, assistive agents, to improve nutrient absorption and bioavailability of nutrients. Essential oil-infused supplements support the digestive process, increase assimilation and bioavailability of nutrients, and enhance immunity. According to Dr. Ashley Cribb, *"Young Living's **oil-infused supplements** are far superior to anything found over the counter at a pharmacy."* (Cribb, 2022).

Table 13: Young Living's Oil-Enhanced Supplements Addressing the Gaps

Product Name	Key Ingredients	Benefits
Golden Turmeric	Turmeric Boswellia resin Ginger	Supports anti-inflammation Bolsters the Immune response Contains prebiotics
Life 9	Contains 9 probiotic strains of bacteria which support overall gut/microbiome health	Aids in healthy digestion Supportive of immunity Daily use supports overall health
Deep Spectra Phyto-nutrition (Black Seed Oil)	Anti-oxidant-and anti-inflammatory rich Black seed and Frankincense Oils	Addresses oxidative stress and inflammation in the body
Master Formula	Full spectrum complex of vitamins, minerals, and food-based nutrients including: turmeric root, cardamom, clove, ginger, & fennel oils B Vitamin Complex	Multivitamin that naturally supports health and wellbeing for the whole body Prebiotic support Addresses oxidative stress Aging requires B vitamins
Mineral Essence	A balanced, full-spectrum ionic mineral complex (60 minerals) Infused with peppermint, cinnamon bark, and lemon oils.	Mineral complex supportive of thyroid gland/metabolism, sleep/emotional support, enzyme building, DNA repair, hormone production, etc.

Multigreens	Blend of spirulina, alfalfa sprouts, barley grass, bee pollen, eleuthero root, kelp Oil blend contains Melissa	Nutrient-dense chlorophyll-based product supportive of overall health and wellness
NingXia Red **Whole Food wolfberry puree with juices of aronia, blueberry, plum, cherry, and pomegranate, infused with citrus and yuzu oils**	Anti-oxidant powerhouse blend providing 18 amino acids, 21 trace minerals, and vitamins B1, B2, B6, and E.	Nutrient-dense antioxidant supplement supportive of overall health and wellness Combats both acute & chronic inflammation. Supportive of eye, digestive, and cardiovascular health
Green Omega 3 Softgels	Marine algal-based formulation, high in DHA and EPA with added flaxseed (ALA) and anti-oxidant-rich clove oil. No fishy aftertaste.	Supports cardiovascular health Supports endocrine & nervous systems Immune support
Super B	Must be ingested daily, as diet is lacking. Necessity for adults.	Supportive of brain health Required as we age
Super C	Vitamin C is fortified with rutin, citrus bioflavonoids, & minerals	Supports immune & respiratory systems, supports overall health
Super Vitamin D **Daily Serving (2 tablets) 50 mcg/2000 IU**	Vitamin D3 combined with lemon balm extract, lime, and Melissa oils	Supports immune, respiratory, skeletal systems Regulates mood and emotions

Embrace the Young Living Lifestyle

For whatever the reason, consumers desire a home that smells good. It is akin to our concept of beauty requiring beauty products. Unfortunately, manufacturers have been more than willing to address this desire by synthetically scenting their products, and in the production of air freshening products. Thus, the 'smell good' home really evolved into a more toxic one, tainted by the presence of synthetic fragrances and phthalates.

Water-based diffusers emit molecules of essential oils carried by droplets of water. For cents on the dollar, the home receives fresh citrus or evergreen-based scents; whatever scent the consumer desires. With pure oils, add several oil droplets to achieve the desired result. During the winter months, increase the number of oil droplets. Another value of using a water-based diffuser, by emitting water into the air, the home becomes humidified. A water-based diffuser acts as a mini-humidifier. Good scents make good sense is a motto to embrace when using essential oils. Earlier in this manuscript, a table was provided presenting the mental/emotional benefits of diffusion of essential oils.

Young Living's optimally produced oils are premium; therefore, value their purity, potency, and efficacy. Essential oils, especially premium, high potency oils provide so many wellness benefits. The antimicrobial nature of essential oils is well documented; consequently, their use (inhalation, topical application, ingestion) in the home is protective. Dozens of research studies have found that essential oils are protective of health by providing relief of symptoms associated with everyday maladies. Diffusion of oils can lower stress levels, provide short-term pain relief, improve mood, open air passage ways, quell food cravings, and relieve digestive discomfort (Lizarraga-Valderrama, 2021).

Oils differ in our responses to them. For example, the scent of peppermint wakes us up, while the scent of lavender, calms us down. According to a recent article in Women's Health Magazine, studies of essential oil scents elicited the following responses. The scent of clary sage reduced blood pressure and breathing rates. The scent of peppermint for men enhanced their performance in the gym. The scent of orange provided to test takers, calmed and reduced symptoms associated with test anxiety. The scent of rosemary left subjects feeling refreshed and mentally stimulated. Drivers, upon inhaling cinnamon bark oil, were more focused and less flustered under difficult driving conditions. The scent of lavender decreased symptoms associated with pre-menstrual syndrome (Rodriguez, 2022). Bottom line, premium essential oils, are protective, assistive, supportive, and corrective. Their antimicrobial nature is protective; while their natural scents work with us to stimulate or calm, as the moment requires. Most importantly, they impart restoration and wellness to a home compromised by a toxic presence.

Young Living's Essential Oils promote a sustainable lifestyle in which toxic exposures are reduced; one that is good for you, good for your family, and good for the Earth. With Young Living, we need to embrace the sustainable practice of **less is more, one swap at a time.**

Table 14: Embrace Less is More, One Swap at a Time, with Young Living

Less Is More One Swap at a Time

Household Use	Toxic Exposure	Young Living's Non-toxic Alternative with Wellness Benefits.
CLEANING	VOCs Formaldehyde preservative Phthalates (EDC) Ammonia Bleach (EDC) Lye Sulfates (SLS, SLES)	THIEVES Line, plant-based, biodegradable ingredients Thieves oil is a blend of clove, cinnamon bark, lemon, *Eucalyptus radiata*, and rosemary oils
PERSONAL CARE	Formaldehyde preservative Parabens (EDC) Phthalates (EDC) Lead & other heavy metals, Diethanolamine (DEA) Sulfates (SLS, SLES)	Personal care products are made with plant-based, biodegradable ingredients and the presence of a diversity of premium, efficacious essential oils. KIDS & Seedlings Line.
NUTRITIONAL SUPPLEMENTS	Quality & Safety Concerns with label reading necessitated Incomplete listing of all ingredients is problematic Companies must use 3rd party testing to insure quality/efficacy	Oil-enhanced to increase absorption and bioavailability of nutrients. Contain high quality, sustainable ingredients.
LIFESTYLE-SUPPORTIVE ESSENTIAL OILS	Phthalate (EDC) exposure via time-release/control-release capsules & pills Promulgation of antibiotic resistance Reliance/dependence on OTC meds and other drugs Pollution source for aquatic species and human drinking water supplies	Multiple research studies support the use of quality essential oils (inhaled, topically applied, or ingested) as being assistive to the body in providing symptomatic relief. The antimicrobial character of oils is protective for the home and its inhabitants.

*youngliving.com/products/2025

As 21st Century consumers, our chemical-based, consumer-oriented society has manufactured a plethora of synthetic toxic chemicals making our household exposures a daily occurrence. Unfortunately, it also has resulted in a glut of synthetics with their toxic presence found in terrestrial and aquatic ecosystems. There is 'no away.' Early in my academic studies, Rachel Carson became my environmental champion, after I read her provocative piece, *Silent Spring*. Provocative because Carson was one of the first to courageously question what were the outcomes of synthetic chemical exposures? Thus, the pointed and eloquent meaning of her title: would Earth's species become so poisoned, that they would be forever silenced? Extinction results in forever silencing.

Unfortunately, Carson's foresight and warnings, unheeded over time, ring true today. We presently exist in Earth's 6th great extinction of catalogued species. Previous extinctions have led to the demise of 80 – 90 percent of Earth's species during each event. The present data numbers are sobering. All species of mammals, birds, reptiles, amphibians, fish, insects, crustaceans, corals, and plants show declining numbers. Human-caused habitat loss, human over-consumption, human population growth, and human pollution, including the presence of toxic chemicals, contribute to this un-paralleled, rapid decline. Scientists estimate that we are losing 10,000 times more species per year than the normal rate (https://www.theworldcounts.com). Perhaps this is of little concern; however, species loss results in a decrease of diversity. Diversity is required for the stability of ecosystems. A diverse ecosystem is more adaptable to withstand change, assuring its sustainability and continued existence over time.

As both an ecologist and biologist, I have no doubt that with the creation and presentation of toxic synthetic chemicals; we have taken hostage so many innocent species. In terms of chemistry and biology, what goes around comes around or what goes in must come out. In other words, none of us lives in isolation from our environment, immediate (home) or otherwise. Earth is a shared planet and there is no planet B. These collective realizations forced me to adopt a different consumer path, not only for the sustainability of my personal home, but also my shared Earthly one. Each of us, as consumers, must make a concerted, mindful, responsible effort to reduce the circuitous path of synthetic chemical toxins, first in our personal homes and with this effort, our Earthly one.

We do not have to be victims of a toxic home or workplace. As a consumer, take responsibility for what is introduced into your personal space. Turn '**this could not be very good**,' into '**this is good**' for me and my family. Adapt, adopt, and embrace a more sustainable lifestyle, such as one exemplified with the use of Young Living Essential Oils and their multiple oil-enhanced product lines. A restorative lifestyle that is **good for you, good for your family, and good for the Earth**.

What Swap Will You Make Today with Young Living Essential Oils?

Table 15: Write In your Young Living Products

Prioritize removal of household toxins, address inadequacies, and replace them with Young Living Essential Oil Products. **Write in below.** ⬇

Household Product	Toxins or Inadequacies to be addressed	Young Living Essential Oils Replacement	Young Living Essential Oils Ingredients
Household Cleaning • Concentrates • Soaps • Detergents • Sanitizers • Toothpastes	**Phthalates (EDC)** **DEA** **SLS, SLES** **Formaldehyde (preservative)** **Bleach (EDC)**	**THIEVES Products:**	**Thieves** essential oil with plant-derived, biodegradable surfactants
Personal Care • Hair Care • Facial Care • Body Care • Lotions • Soaps/gels • Creams	**Phthalates (EDC)** **Parabens (EDC)** **Formaldehyde** **DEA** **SLS, SLES** **Oxybenzone (EDC)** **Heavy Metals**	**Personal Care Products**	Presence of premium, efficacious essential oils with plant-derived, biodegradable ingredients
Child Care Products found in household location • Nursery • Bathroom • Laundry Room • Kitchen • Outdoors (sun-protective & pest deterrent)	**Phthalates (EDC)** **Parabens (EDC)** **DEA** **SLS, SLES** **Oxybenzone (EDC)** **DEET** **Heavy Metals**	**KidScents & SEEDLINGS** Lines	Presence of premium, efficacious essential oils with plant-derived, biodegradable ingredients Hypoallergenic Easy application

Nutrition & Foods • Gut support • Anti-oxidant support • Immune-supportive • Enhanced bioactivity • Enhanced digestion	Nutritional gaps occur with a nutrient-deficient food supply and diet due to processed foods, aging, picky eaters. **BPA** **Parabens (EDC)** **Dyes/flavors**	**Nutritional Supplements**	Presence of premium, efficacious oils enhance bioactivity and nutritional uptake of supplements Bolster nutritional support Aging-supportive Support digestion
Other Household Uses • Air purification • OTC medications • Mindfulness support	**Phthalates (EDC)** **Formaldehyde** **DEET** **Pesticides** **Herbicides** **Oxybenzone (EDC)**	**Other Household Products** Water and Oil-based diffusers Inhibits pests Inhibits UV rays	Diversity of premium, efficacious essential oils with plant-derived, biodegradable ingredients Antimicrobial Anti-inflammatory Naturally analgesic

References

Abd El-Hack. Mohamed E. et al. "Essential oils and their nanoemulsions as green alternatives to antibiotics in poultry nutrition: a comprehensive review." Poultry Science. 101(2): February 2022. https://doi.org?10.1016/j.psj.2021.101584).

Ahern, Thomas, et al. "Medication-Associated Phthalate Exposure and Childhood Cancer Incidence." J.Natl Cancer Inst. 2022 Jun 13;114(6): 885-894.doi:10.1093/jnci/djac045.

Ajmera, Rachael, MS, RD. "What Are the Health Benefits of Black Seed Oil? Dec. 18, 2024. Healthline.com/health/food-nutrition/black-seed-oil-benefits.

Akanbi, CA, "Endocrine-disrupting chemicals (EDCs) and epigenetic regulation in embryonic development: Mechanisms, impacts, and emerging trends." Toxicology Reports. June 2025 14(1):101885 DOI: 10.1016/j.toxrep.2024.101885.

Alammar. N. et al. "The impact of peppermint oil on the irritable bowel syndrome; a meta-analysis of the pooled clinical data." BMC Complement Altern Med 19, 21(2019). https://doi.org/10.1186/s12906-018-2409-0.

Albrecht, Rita. "If drinking water can be bad in Bemidji, it can be contaminated anywhere." Minnsotareformer.com/2022/02/11/if-drinking-water-can-go-bad-in-Bemidji-it-can-become-contaminated-anywhere-opinion/

American Society for Biochemistry and Molecular Biology. "Thyme oil can inhibit COX2 and suppress inflammation." ScienceDaily, 14 January 2010. www.sciencedaily.com/releases/2010/01/100113122306.htm.

Amos, Owen. "Why opioids are such an American problem." 25 October 2017. BBC News, Washington DC. bbc.com/news/world-us-canada-41701718.

Arrigo, Federica., et al. "Phthalates and their effects on human health: Focus on erythrocytes and the reproductive system." Comparative Biochemistry and Physiology Part C: Toxicology & Pharmacology. Volume 270, August 2023, 109645. https://doi.org/10.1016/j.cbpc.2023.109645.

ATSDR. "How Are Newborns, Infants, and Toddlers Exposed to and Affected by Toxicants?" May 25, 2023. Archive.cdc.gov/www_atsdr_cdc_gov/csem/pediatric-environmental-health/newborns_infants_toddlers.html.

Bansal, Bhavika. "Unveiling Americans' Attitudes Towards Over-The-Counter Drugs." American Pharmaceutical Review. Innovations at Interphex 2024 Supplement. March, 11, 2024. pgs.18-19.

Barrows, Julie N., et al. "Color Additives History." Food Safety Magazine. October/November 2003 Issue. Fda.gov/industry/color-additives/color-additives-history

Bassi, Margherita. Gizmodo.com/a-new-report-suggests-we-all-may-be-brushing-our - teeth-with-lead-and-other-toxic-metals-2000591000.

Beck, Echardt C. "The Love Canal Tragedy." EPA Journal-January 1979. Epa.gov/archive/epa/aboutepa/love-canal-tragedy,html.

Bo Cao, et al. "Seeing the Unseen of the Combination of Two Natural Resins, Frankincense and Myrrh: Changes in Chemical Constituents and Pharmacological Activities." Molecules. 2019 Aug 24;24(17): 3076.doi:10,3390/molecules24173076.

Bongers, F, et al. "Frankincense in peril." Nature Sustain 2, 602-610 (2019) hhttps://doi.org/10.1038/s41893-019-0322-2.

Buck-Louis, Germaine M. et al. "Lifestyle and pregnancy loss in a contemporary cohort of women recruited before conception: The LIFE Study." Fertility and Sterility, Volume 106, Issue 1, 180-188. July 2016.

Caputo, Ludmila de Souza., et al. "Copaiba oil minimizes inflammation and promotes parenchyma reepithelization in acute allergic asthma model induced by ovalbumin in BALB/c mice." Front Pharmacol.2024 Apr 11;15: 1356598.doi:10.3389/fphar.2024.1356598.

Chávez-Delgado, Emily, and Daniel A Jacobo-Velázquez. "Essential Oils: Recent Advances on Their Dual Role as Food Preservatives and Nutraceuticals against the Metabolic Syndrome." Foods. 2023 Mar 3;12(5): 1079.doi:10.3390/foods12051079.

Cohen, Sandy. "If you want to boost immunity, look to the gut." March 19, 2021. uclahealth.org/news/article/want-to-boost-immunity-look-to-the-gut.

Cooper, Richard J. "Over-the-counter medicine abuse-a review of the literature." J Subst Use. 2013/04/01 PubMed. DOI:10.3109/14659891.2011.615002.

Cribb, Ashley H. Supplement Field Guide. 2022. Growing Healthy Homes LLC. ISBN: 978-1-7370846-4-8.

Cronkleton, Emily. "Essential Oils for Pain Relief: Does it Work?" March 31 2022.

https://www.healthline.com/health/pain-relief/essential-oils-for-pain//takeaway.

Davis, Donald R., et al. "Changes in USDA food composition data for 43 garden crops, 1950-1999." J Am Coll Nutr. 2004 Dec;23(6):669-82. Doi:10.1080/07315724.2004.10719409.

"Deaths linked to 'hormone disruptor' chemical costs billions in lost US productivity. October 12, 2021. Sciencedaily.com/releases/2021/10/211012080113.htm.

DeNicola, Nathaniel., et al. "Toxic environmental exposures in maternal, fetal, and reproductive health." The Medical Society Consortium on Climate & Health. Contemporary OB/GYN Sep 12, 2018. Medsocietiesforclimatehealth.org/members-in-action/toxid-environmental-exposures-maternal-fetal-reproductive-health/.

DiNardo, Joseph and Craig Downs. "Dermatological and environmental toxicological impact of the sunscreen ingredient oxybenzone/benzophenone-3." J Cosmet Dermatol. 2018 Feb;17(1):15-19. Doi:10.1111/jocd.12449. Epub 2017 Oct 31.

de Sousa, DP, et al. "Essential oils: Chemistry and Pharmacological Activities." Biomolecules. 2023 Jul 18;13(7):1144. doi: 10.3390/biom13071144.

"Drugs in the Water." June 1 2011. Health.harvard.edu/newsletter_article/drugs-in-the-water.

Ducharme, Jaime. "RFK Jr. Wants to Get Rid of Artificial Food Dyes-How Healthy Are the Alternatives?" April 30,2025. health.com/natural-food-dye-alternatives-health-11724251.

Duty, Susan, et al. "Personal care product use predicts urinary concentrations of some phthalate monoesters." Environ Health Perspectives. 113(11):1530-5. DOI:10.1289/ehp.8083.

Endocrine-Disrupting Chemicals in the Home. 2021. www.madesafe.org/pages/reports-guides.

Engdahl, Elin, et al. "DNA methylation at GRIN2B partially mediates the association between prenatal bisphenol F exposure and cognitive functions in 7-year-old children in the SELMA study." Environ Int. 2021 Nov:156: 106617.doi:10.1016/j.envint.2021.106617. Epub 2021 May 18.

Engel, Stephanie, et al. "Neurotoxicity of Otho-Phthalates: Recommendation for Critical Policy Reforms to Protect Brain Development in Children." American Journal of Public Health (AJPH). April 2021. Published Online: March 10, 2021.

EPA.gov/pcbs. May 30, 2025.

Essential Oil Desk Reference. 7th ed. Life Science Publishing. 2016.https: www.discoverlsp.com

"EWG Cleaners Hall of Shame Reveals Hidden Hazards." ewg.org/news-insights/news-release/ewg-cleaners-hall-shame-reveals-hidden-hazards. April 23.2012.

Fast, Austin and Ignacio Calderon. "EPA will roll back limits of 4 'forever chemicals,' See is they were found in your water." USA Today May 15, 2025.

fda.gov/cosmetics/cosmetic-products/limiting-lead-lipstick-and-other-cosmetics.11/22/2022.

Feldscher, Karen. "Exposure to phthalates may raise risk of pregnancy loss, gestational diabetes." November 8, 2016. Hsph.harvard.edu/news/phthalates-exposure-pregnancy-loss-gestational-diabetes/.

"Five Consumer-Facing Challenges of the OTC Medicine Experience." Council Post. Dec. 26, 2019. Forbes.com/councils/theyec/2019/12/26/five-consumer-facing-challenges-of-the-otc-medicine-experience/.

Fuller, R. D., Ph.D. State University of New York (Plattsburgh). Retired. Personal communication. 2025.

Geya Sree, Chendruru, et al. "Phthalate toxicity mechanisms: An update." Comparative Biochemistry and Physiology Part C: Toxicology & Pharmacology 2023 Jan:263: 109498.doi: 10.1016/j.cbpc.2022.109498. Epub 2022 Oct 28.

Ghlichloo, Ida and Valerie Gerriets. "Nonsteroidal Anti-Inflammatory Drugs (NSAIDs" May 1 2023.ncbi.nimnih.gov/books/NBK547742/.

Green Cleaning, Sanitizing, and Disinfecting: A Toolkit for Early Care and Education Fact Sheet. https://wspehsu.ucsf.edu/projects/environmental-health-in-early-care-and-education-project/.

Green, Mark P., et al. "Endocrine disrupting chemicals: Impacts on human fertility and fecundity during the peri-conception period." Environmental Research Vol.194. March 2021, 110694. https://doi.org/10.1016/j.envres.2020.110694.

Greiner, Peter, et al. "Occurrence of perchlorate in sodium hypochlorite." JournalAWWA. 01 November 2008. https://doi.org/10.1002/j.1551-8833.2008.tb09773.

Gunnars, Kris. "10 Health Benefits of Turmeric and Curcumin." November 27, 2023. https://www.healthline.com/top-10-evidence-based-health-benefits-of-tumeric#bottom-line).

Harada, Hiroki, et al. "Linalool odor-induced anxiolytic effects in mice." Front. Behav. Neurosci. 2018 Oct 23:12:241. Doi:10.3389/fnbeh.2018.00241. eCollection 2018.

Hardin, Tod. "3M Agrees to $10.3 Billion Settlement in PFAS-Related Lawsuits," dcreport.org/2023/06/25/3m-agrees-to-10-3-billion-settlement-in-pfas-related-lawsuits/..

Harley, Kim G. "Reducing Phthalate, Paraben and Phenol Exposure from Personal Care Products in Adolescent Girls: Findings from the HERMOSA Intervention Study." Environmental Health Perspectives. 2016 Oct;124(10): 1600-1607.doi:10.1289/ehp.1510514. Epub Mar 7.

Hilal, Bisma, et al. "Recent advancements in deciphering the therapeutic properties of plant secondary metabolites: phenolics, terpenes, and alkaloids." Plant Physiology and Biochemistry. Volume 211, June 2024, 108674. https://doi.org/10.1016/j.plaphy.2024.108674.

Houlihan, Jane, et al. "Not too pretty." July 8, 2002. Static.ewg.org/reports/2002/NotToopretty.

https://www.the worldcounts.com/challenges/planet-earth/forests-and-deserts-/species-extinction-rate/story.

Hu, Guo-Xin., et al. "Phthalate-induced testicular dysgenesis syndrome: Leydig cell influence." Trends Endocrinol Metab. 2009 Mar 9;20(3):139-145. Doi: 10.1016/j.tem.2008.12.001.

"Impact of EDCs on Metabolism and Obesity." Endocrine Society. Endocrine.org/topics/edc/what-edcs-are/common-edcs/metabolic#.

Karimi, Nasrin., et al. "The Effect of Aromatherapy with *Pelargonium graveolens* on the Fatigue and Sleep Quality of Critical Care Nurses during the Covid-19 Pandemic: A Randomized Controlled Trial. Explore (NY). 2023 June 20. Online ahead of print.doi: 10.1016/j.explore.2023.06.006.

Khairnar, Rutuja P., et al. "Rosmarinus Officinalis L.: Used for the Treatment of Hair Loss.: Int. J. in Pharm. Sci., 2023, Vol 1, Issue 12, 524-533. Review.

Kim, In-Hee, et al. "Essential oil inhalation on blood pressure and salivary cortisol levels in prehypertensive and hypertensive subjects." Evid Based Complement Alternat Med. 2012:2012: 984203.doi:10.1155/2012/984203. Epub 2012 Nov 19

LaMotte, Sandee. "Common household plastics linked to thousands of global deaths from heart disease." CNN Health and Wellness. April 30, 2025 cnn.com/2025/04/29/health/phthalates-heart-disease-wellness.

Liu, Sa, et al. "Concentrations and Potential Health Risks of Metals in Lip Products." Environ, Health Perspectives. 2 May 2013. 121:6 705-710.

Lizarraga-Valderrama, Lorena R. "Effects of essential oils on central nervous system: Focus on mental health." Phytother Res. 2021 Feb;35(2):657-679. Doi:10.1002/ptr.6854. Epub 2020 Aug 29.

Mandai, Ananya. "Links between deodorants and breast cancer refuted but more information needed: Study." Jan 16, 2012. News-medical.net/news/20120116/Links-between-deodorants-and-breast-cancer-refuted-but-more-information-needed-study.

Markowitz, Gerald. "From industrial toxins to worldwide pollutants: a brief history of polychlorinated biphenyls." Public Health Rep. 2018 Sept;133(6): 721-725.doi: 10.1177/0033354918801578.Epub 2018 Sep28.

Marohn, Kirsti. "Forever chemicals leave costly water problem in Bemidji, cities across the country." mprnews.org/story2019/02/14/pfas-leaves-costly-water-problem-in-bemidji-and-other-cities.

Matsubara E. et al. "Bornyl acetate induces autonomic relaxation and reduces arousal level after visual display terminal work without any influences of task performance in low-dose condition" Biomed Res. 2011 Apr;32(2): 151-7.doi: 10.2220/biomedres.32.151.PMID:21551951.

Matsumoto, Tamaki, et al. "Effects of olfactory stimulation from the fragrance of the Japanese citrus fruit Yuzu on mood states and salivary chromogranin A as an endocrinologic stress marker." J Altern Complement Med. 2014 June 1;20(6): 500-506.doi:10.1089/acm.2013.0425.

McConnell, Tatum. "Stay safe this summer: Ditch sunscreens with "troublesome" oxybenzone, experts say." May 23, 2023. Environmental Health News. Ehn.org/reef-safe-sunscreen.

Miller, Korin. "This food was just linked to heart disease and a serious brain condition." May 22, 2025. Prevention.com/health/a64829081/ultra-processed-food-heart-disease-parkinsons-study/.

Murray, Rheana. "You'll spend as much money on makeup as it costs to buy a house in your lifetime." March 30, 2017. Today.com/style/women-spend-200k-makeup-lifetime-skinstore-com-survey-t109772.

Nan Lv, Xiao, et al. "Aromatherapy and the central nerve system (CNS): therapeutic mechanisms and its associated genes." Curr Drug Targets. 2013 Jul;14(8):872-9. DOI:10.2174/13894501113140800007.

National Center for Biotechnology Information. "PubChem Compound Summary for CID 3034034, Quinine" PubChem, https://pubchem.ncbi.nim.nih.gov/compound/Quinine. Accessed 11 June, 2025. https://doi.org/10.1038/s41591-024-03453-1.

Niehs.nih.gov/health/topics/agents/endocrine.

Nihart, Alexander J. et al. "Bioaccumulation of microplastics in decedent human brains." Nat Med 31, 1114-1119 (2025).

Ninkuu, Vincent, et al. "Biochemistry of Terpenes and Recent Advances in Plant Protection.: Int J Mol Sci. 2021 May 27;22(11): 5710.doi: 10.3390/jms22115710.

NJ.gov/health/eoh/rtkweb/documents/fs/0686.pdf.

NOAA. "Skincare Chemical and Coral Reefs." June 20.2024. https://oceanservice.noaa.gov/news/sunscreens-corals.html

Ods.od.nih.gov/factsheets/WYNTK-Consumer.

Oh, Jiwon, et al. "Exposures to Contemporary and Emerging Chemicals among Children Aged 2 to 4 Years in the United States Environmental Influences on the Child Health Outcome (ECHO) Cohort." Environmental Science & Technology, 2025; DOI: 10.1021/acs.est.4c13605.

Pattnaik, S, et al. "Antibacterial and antifungal activity of ten essential oils in vitro.: Microbios. 1996;86(349):237-46. PMID:8893526.

Petersen, Dorene. "Ten Essential Oils that Naturally Repel Insects," July 2024. Aches.edu/blog/green-cleaning-repel-insects/.

Pezantes-Orellana, Camilla, et al. "Essential oils: a systematic review on revolutionizing health, nutrition, and omics for optimal well-being." Front Med (Lausanne). 2024 Feb 16:11: 1337785.doi:10.3389/fmed.2024.1337785.

Price, Kiley. "Traces of pharmaceuticals are contaminating rivers and oceans worldwide." May 15, 2025. The Daily Climate. Dailyclimate.org/traces-of-pharmaceuticals-are-contaminating-rivers-and-oceans-worldwide-2672023703.html.

"Product Profile: Bug Repellent." July 6, 2023. madesafe.org/blogs/viewpoint/product-profile-bug-repellent.

Rashidi-Fakari, Fahimeh, et al. "The effect of aromatherapy by essential oil of orange on anxiety during labor: A randomized clinical trial." Iran J. Nurs Midwifery Res. 2015 Nov-Dec;20(6):661-4. Doi: 10.4103/1735-9066.170001.

Reynolds, Pam. "Exposure to Chemicals is Making Our Kids Sick." April 8, 2025. clf.org/blog/exposure-to-chemicals-is-making-our-kids-sick 2025.

Rodriguez, Tori., et al. "The 15 Best Essential Oils for Aromatherapy and How to Choose and Use Them." July 23, 2022. womenshealthmag.com/health/a19904702/essential-oils/.

Rudel, Ruthann, et al. "Semivolatile Endocrine-Disrupting Compounds in Paired Indoor and Outdoor Air in Two Northern California Communities." Environ Science & Technology Vol 44/Issue 17:6523-6904. August 3, 2010.

safecosmetics.org/chemicals/lead-and-other-heavy-metals/

Salafard, Mahla., et al. "The Effect of Orange Essential Oil Aromatherapy on Sleep Quality in Hospitalized Children." Preventative Care in Nursing and Midwifery Journal. 13(1):83-90. Nmcjournal.zums.ac.ir/article-1-824-en.html.

Sattayakhom, Apsorn, et al. "The effects of essential oils on the nervous system: a scoping review.: Molecules 2023 Apr 27;28(9) 3771.doi:10.3390/molecules28093771.

Schaefer, Anna. "What is Sodium Lauryl Sulfate (SLS)?" May 23 2023. healthline.com/health/beauty-skin-care/what-is-sodium-lauryl-sulfate/.

Sharma, Mansi, et al. "Essential oils as anticancer agents: potential role in malignancies, drug delivery mechanisms, and immune system enhancement." Biomedicine & Pharmacotherapy Volume 146, Feb 2022. 112514 https://doi.org/10.1016/j.biopha.2021.112513.

Smith, Andrew. "Hidden in Plain Sight: The Chemical Burden of Personal Care Products.: March 5, 2025. Rutgers.edu/news/hidden-plain-sight-chemical-burden-personal-care-products.

Sparks Akers, Alicia. November 13, 2024 "Is it safe to eat essential oils?" medicalnewstoday.com/articles/can-you-eat-essential-oils.

Stoiber, Tasha. "What are parabens, and why don't they belong in cosmetics?" April 9, 2019. ewg.org/what-are-parabens.

Suzuki, David. "The Dirty Dozen: Coal Tar Dyes." 2025. Davidsuzuki.org/living-green/the-dirty-dozen-coal-tar-dyes/.

Swan, Shanna, and Stacey Colino. Count Down: Our Modern World Is Threatening Sperm Counts, Altering Male and Female Reproductive Development, and Imperiling the Future of the Human Race. New York, New York. Scribner. February 23, 2021.

Tariq, Saika, et al. "A comprehensive review of the antibacterial, antifungal, and antiviral potential of essential oils and their chemical constituents against drug-resistant microbial pathogens.: Microbial Pathogenesis. 134: September 2019, 103580. https://doi.org//10.1016/j.micpath.2019.103580.

"Toxic Matters: New and Improved." 2013. prhe.ucsf.edu/toxic-matters

"Toxicological Profile for Formaldehyde." ncbi.nim.gov/books/NBK597627/.

Uzuncakmak, Tugba and Sultan Ayaz Alkaya. "Effect of aromatherapy on coping with premenstrual syndrome: A randomized controlled trial." Complementary Therapies in Medicine. Volume 36, Feb 2018, pgs. 63-67. https://doi.org/10.1016/j.ctim.2017.11.022.

Venkatesan, Krishnaraju., et al. "Protective Effects of Frankincense Oil on Would Healing: Downregulating Caspase-3 Expression to Facilitate the Transition from the Inflammatory to Proliferative Phase." Pharmaceuticals (Basel). 2025 Mar 13;18(3): 407.doi:10.3390/ph18030407.

Wambolt, Cadee. "Exposure to toxic chemicals in consumer products in the United States." January 2021. www.ballardbriefs.org.

Wang, Yufei, and Haifeng Qian. "Phthalates and Their Impacts on Human Health." Healthcare (Basel). 2021 May 18;9(5): 603.doi: 10.3390/healthcare9050603.

Ward, et al. "Childhood and adolescent cancer statistics." CA Journal for Clinicians. 214:64(2):83-103. 2014.

Westervelt, Amy. "Phthalates are everywhere, and the health risks are worrying. How bad are they really?" The Guardian. 2015.2/10. (the guardian.com/lifeandstyle/2015/feb/10/phthaltes-plasticizers-chemicals-research-analysis).

"What is BPA, and what are the concerns about BPA?" March 24, 2023. mayoclinic.org/healthy-lifestyle/nutrition-and-healthy-eating/expert-answers/bpa.

www.youngliving.com/us/farms/2025.

www.youngliving.com/products/2025.

Yada, Yukihiro, et al. "Overseas survey of the effect of cedrol on the autonomic nervous system in three countries." Clinical trial. J Physiol Anthropol. 2007 May; 26(3):349-54. Doi:10.2114/jpa2.26.349.

Zahra, Heydari., et al. "Effects of Rosemary Oil on Osteoarthritis Symptoms in the Elderly: A Double-Blind Randomized Control.: J Occup Health Epidemiol 2022, 11(3):171-179. Johe.rums.as.ir/article-1-544-en.html.

About the Author

Katherine T. Fuller (Kathie) retired as a public-school teacher in 2013. Her career spanned 33-years in the classroom. Subjects taught included biology, forensics science, science of foods & cooking, and AP Biology. She continued to teach at the college level, as a lecturer (BIO 101) at Plattsburgh State University, NY. After his retirement as a college professor in 2017, husband, Bob, and Kathie, moved from New York to Bemidji, MN. They now live on a 40-acre farm where Kathie grows vegetables, herbs, and flowers for sale from her greenhouse. They consume organically-grown vegetables and herbs from her garden. They also maintain a winter home in Litchfield Park, AZ.

Kathie, influenced by her great friend and mentor, Connie McDanel, embraced a Young Living Essential Oils lifestyle over two decades ago. Her daughter, Rachel's asthmatic nature, induced by bleach compounds serving as an asthmatic trigger, promoted the search for less toxic household cleaning and personal care products. Once adopted, she practiced the Young Living Essential Oil lifestyle with Bob and her two children, Russell and Rachel. They, in turn, are now imparting the use of diffusers and oils, topical application of oils, and ingestion of oil-enhanced nutritional supplements with Kathie's grandchildren, Ronan, and Gerry (Russ and Mary Grace), and Lucy, Theo, and Bobby (Rachel and Tim).

Kathie's degrees in biology (BS) and ecology (MS) have influenced her minimalistic lifestyle approach evidenced in this manuscript. 'Less is More' aligns with the concept of sustainability which values and respects our environment and its resources. She considers its conservation paramount for survival of Earth's inhabitants. **Our children and our grandchildren are our legacy. They are what we leave behind and at the same time what we leave ahead. Embracing and practicing sustainability, exemplified with the adoption of a Young Living Essential Oil lifestyle, is good for my husband and I, our children, and most importantly, our grandchildren.**

Made in the USA
Las Vegas, NV
02 November 2026

57142257R00052